Control, Simulation, and Monitoring of Thermal Processes in Power Plants

Control, Simulation, and Monitoring of Thermal Processes in Power Plants

Editor

Pawel Madejski

MDPI • Basel • Beijing • Wuhan • Barcelona • Belgrade • Manchester • Tokyo • Cluj • Tianjin

Editor
Pawel Madejski
AGH University of Science
and Technology
Poland

Editorial Office
MDPI
St. Alban-Anlage 66
4052 Basel, Switzerland

This is a reprint of articles from the Special Issue published online in the open access journal *Energies* (ISSN 1996-1073) (available at: https://www.mdpi.com/journal/energies/special_issues/Thermal_Processes_in_Power_Plants).

For citation purposes, cite each article independently as indicated on the article page online and as indicated below:

LastName, A.A.; LastName, B.B.; LastName, C.C. Article Title. *Journal Name* **Year**, *Volume Number*, Page Range.

ISBN 978-3-0365-6796-9 (Hbk)
ISBN 978-3-0365-6797-6 (PDF)

Cover image courtesy of Paweł Madejski

Contents

About the Editor . . . . **vii**

Paweł Madejski, Piotr Michalak, Michał Karch, Tomasz Kuś and Krzysztof Banasiak
Monitoring of Thermal and Flow Processes in the Two-Phase Spray-Ejector Condenser for
Thermal Power Plant Applications
Reprinted from: *Energies* **2022**, *15*, 7151, doi:10.3390/en15197151 . . . . **1**

Zhiling Luo and Qi Yao
Multi-Model-Based Predictive Control for Divisional Regulation in the Direct Air-Cooling
Condenser
Reprinted from: *Energies* **2022**, *15*, 4803, doi:10.3390/en15134803 . . . . **23**

Goran S. Kvascev and Zeljko M. Djurovic
Water Level Control in the Thermal Power Plant Steam Separator Based on New PID Tuning
Method for Integrating Processes
Reprinted from: *Energies* **2022**, *15*, 6310, doi:10.3390/en15176310 . . . . **41**

Paweł Madejski, Tomasz Kuś, Piotr Michalak, Michał Karch and Navaneethan Subramanian
Direct Contact Condensers: A Comprehensive Review of Experimental and Numerical
Investigations on Direct-Contact Condensation
Reprinted from: *Energies* **2022**, *15*, 9312, doi:10.3390/en15249312 . . . . **59**

**Paweł Ziółkowski, Paweł Madejski, Milad Amiri, Tomasz Kuś, Kamil Stasiak,
Navaneethan Subramanian, et al.**
Thermodynamic Analysis of Negative CO_2 Emission Power Plant Using Aspen Plus, Aspen
Hysys, and Ebsilon Software
Reprinted from: *Energies* **2021**, *14*, 6304, doi:10.3390/en14196304 . . . . **91**

**Vishwajeet, Halina Pawlak-Kruczek, Marcin Baranowski, Michał Czerep,
Artur Chorążyczewski, Krystian Krochmalny, et al.**
Entrained Flow Plasma Gasification of Sewage Sludge–Proof-of-Concept and Fate of Inorganics
Reprinted from: *Energies* **2022**, *15*, 1948, doi:10.3390/en15051948 . . . . **119**

Paweł Madejski, Karolina Chmiel, Navaneethan Subramanian and Tomasz Kuś
Methods and Techniques for CO_2 Capture: Review of Potential Solutions and Applications in
Modern Energy Technologies
Reprinted from: *Energies* **2022**, *15*, 887, doi:10.3390/en15030887 . . . . **133**

Maciej Bujalski and Paweł Madejski
Forecasting of Heat Production in Combined Heat and Power Plants Using Generalized
Additive Models
Reprinted from: *Energies* **2021**, *14*, 2331, doi:10.3390/en14082331 . . . . **155**

About the Editor

Paweł Madejski

Paweł Madejski DSc, PhD, Eng. is Associate Professor at AGH University of Science and Technology in Kraków, Poland. In the years 2010–2018, he worked as a Senior Specialist for Measurement I&C Research and for Thermodynamic Research in the Research and Development Department of the energy group EDF Polska SA (later PGE SA). In 2017, he became a laureate of the Leader of Engineering under 40 competition organized by Control Engineering Polska magazine. His scientific interests are mainly focused on research on the efficiency of flow-thermal processes and combustion processes in energy technologies, monitoring, modeling, and simulating the operation of power units and steam boilers, and CFD modeling of the operation of power machines and devices. Currently, he lectures on energy technologies, thermodynamics, steam boilers, power plants, and combined heat and power plants at AGH University of Science and Technology in Kraków. He is author and co-author of over 100 research papers (including articles, book chapters, monographs, and reviewed conference materials). He received his MSc (2009) and PhD degree (2014) in the field of construction and exploitation of machinery and his post-doctoral degree of doctor habilitated (2020) conferment in the field of engineering and technical sciences in the discipline of mechanical engineering.

energies

Article

Monitoring of Thermal and Flow Processes in the Two-Phase Spray-Ejector Condenser for Thermal Power Plant Applications

Paweł Madejski [1], Piotr Michalak [1,*], Michał Karch [1], Tomasz Kuś [1] and Krzysztof Banasiak [2]

[1] Department of Power Systems and Environmental Protection Facilities, Faculty of Mechanical Engineering and Robotics, AGH University of Science and Technology, Al. Mickiewicza 30, 30-059 Kraków, Poland
[2] SINTEF Energy, 7034 Trondheim, Norway
* Correspondence: pmichal@agh.edu.pl

Abstract: The paper deals with the problem of accurate measuring techniques and experimental research methods for performance evaluation of direct contact jet-type flow condensers. The nominal conditions and range of temperature, pressure and flow rate in all characteristic points of novel test rig installation were calculated using the developed model. Next, the devices for measurement of temperature, pressure and flow rate in a novel test rig designed for testing the two-phase flow spray ejector condensers system (SEC) were studied. The SEC can find application in gas power cycles as the device dedicated to condensing steam in exhaust gases without decreasing or even increasing exhaust gas pressure. The paper presents the design assumptions of the test rig, its layout and results of simulations of characteristic points using developed test rig models. Based on the initial thermal and flow conditions, the main assumptions for thermal and flow process monitoring were formulated. Then, the discussion on commercially available measurement solutions was presented. The basic technical parameters of available sensors and devices were given, discussed with details.

Keywords: direct contact condenser; spray-ejector condenser; mass flow measurement; experimental test rig

Citation: Madejski, P.; Michalak, P.; Karch, M.; Kuś, T.; Banasiak, K. Monitoring of Thermal and Flow Processes in the Two-Phase Spray-Ejector Condenser for Thermal Power Plant Applications. *Energies* **2022**, *15*, 7151. https://doi.org/10.3390/en15197151

Academic Editor: Ron Zevenhoven

Received: 29 July 2022
Accepted: 26 September 2022
Published: 28 September 2022

Publisher's Note: MDPI stays neutral with regard to jurisdictional claims in published maps and institutional affiliations.

1. Introduction

Due to the growing importance of heat and electricity generation efficiency in thermal power plants, while taking into account recent political determinants and resulting legal regulations, especially on CO_2 emissions, various solutions have been developed in recent years.

Ziembicki et al. [1] analysed heat recovery from combustion gases in heating plants using gaseous and liquid fuels. Tic and Guziałowska-Tic [2] investigated the cost-efficiency dependence of the combustion of solid fuel improvement in power boilers, especially fine coal.

Apart from these efficiency-related measures, techniques are also used aiming at a reduction in CO_2 emissions [3,4]. From them, an important role plays carbon capture and storage (CCS) utilising pre-combustion, post-combustion or oxyfuel combustion approaches [5,6].

A pure CO_2 stream from the combustion gases, when combusting the carbonaceous fuel in thermal power plants, can be achieved involving post-combustion and oxy-combustion processes [7].

In [8], authors presented the performance analysis of a designed advanced ultra-supercritical (A-USC) coal-fired 700 MW power plant. To reduce CO_2 emissions, the post-combustion carbon capture and storage (CCS) unit, with a CO_2 removal rate of 90%, was considered. The calculated net efficiency of an A-USC unit was 47.6% and 36.8% with and without CCS, respectively. The electricity output penalty was 362.3 kWh_{el}/tCO_2. However, application of the oxy-combustion method may result in a lower penalty, around 4% [9].

An interesting concept is a power plant based on biomass, which has zero net emission and is connected with CCS and results in negative emissions [7]. Further development of this idea for sewage sludge in combination with oxy-combustion and CO_2 capture was presented in [10–12]. The whole system consists of three basic parts. In the first of them, sewage sludge is processed into syngas. Next, the resulting fuel is burned in pure oxygen in a special wet combustion chamber. Finally, produced hot steam and carbon dioxide are supplied to a gas turbine cooperating with a spray ejector condenser (SEC) with CO_2 capture. The main task of SEC is to condense the water vapour from the exhaust gases while creating a compact system structure.

Spray condensers (spray ejector condensers, direct contact condensers, jet condensers) have been used in industry since the beginning of the 20th century [13], covering a wide range of various applications in chemical engineering, water desalination, air conditioning and energy conversion processes.

Regarding the latter, in power plants, SECs were used to improve performance and to obtain lower operating costs, and numerous studies on these issues were presented recently.

Desideri and Di Maria [14] simulated the impact of the geothermal fluid main characteristics on the performance of 20 and 60 MW geothermal power plants. Direct contact condensers were used to condense and separate geothermal steam from the non-condensable gases. In [15], waste heat from a thermoelectric power plant was utilised to produce fresh water in a diffusion driven desalination facility using a direct contact condenser. The saturated air and vapour mixture is cooled and dehumidified by the water inside the condenser. Another application of SEC was a cooling cycle in a combined cooling, heat and power (CCHP) system based on a micro-steam turbine and devoted to a residential building [16]. Marugán-Cruz et al. [17] modelled a cooling system involving a direct contact jet condenser to reduce the water consumption of a concentrated solar tower power plant. In [18], authors analysed steam consumption of conventional lignite pre-drying systems when using heat recovery from the process of wet flue gas desulphurisation (WFGD). The WFGD tower was used as a direct-contact heat exchanger and the source of heat for the first pre-drying stage. Zhao et al. [19] simulated heat recovery from flue gas in a gas fired CHP plant with a steam driven absorption heat pump and a direct contact heat exchange tower. In [20], authors applied the O_2/CO_2 cycle to a natural gas combined cycle (NGCC) using a gas turbine. The required oxygen was produced by an air separation unit including a spray condenser. Simulations in Aspen Plus resulted in a net efficiency loss of 8.1%-points when applying separation of CO_2.Garlapalli et al. [21] analysed CO_2 capture from a coal-fired power plant. To reduce the reboiler energy consumption, they proposed a new solution for heat recovery from the flue gas, using a direct contact heat exchanger. The latter was also involved in post-combustion CO_2 capture.

Contrary to numerous simulation works covering a wide range of direct contact condensers, only a limited number of case studies were based on in situ measurements. Wei et. al. [22] analysed a flue gas heat recovery system in a coal-fired CHP. They applied heat recovery with direct-contact cooling with low-temperature water from the heat pump. This heat was used then in a heating network. In [23], authors investigated the utilisation of available excess steam in a 200 MW geothermal power plant with a dry steam cycle. The spray ejector condenser was used to condense the water vapour from the exhaust gases. The simulation model was then validated against operational data and optimisation of the power plant was carried out.

There is a lack of laboratory scale test rigs devoted to measurements for a spray condenser for application in CO_2 capture. Papers were presented on various applications of direct contact condensers in thermal engineering, as pyrolysis of biomass [24], refrigeration cycle [25,26] or directed into construction details, performance characteristics or transient behaviour of SEC [27–30].Therefore, there is a need to experimentally investigate an application of a spray condenser in a CO_2 capture system in a thermal power plant cycle. From a practical point of view, the choice of measurement methods and instruments is important. This issue was also poorly represented in the publications presented. Authors mainly focused on the description of the equipment used together with values of measurement errors. This is the research gap that defines the aim of the paper.

The next section briefly presents the main assumptions for the conceptual design of a test rig developed within the scope of the project for a negative CO_2 emission gas power plant [31–33]. The circuit layout and preliminary simulations are also described. Then, the variability ranges of thermal and flow parameters in the selected points of the test rig are given. This way, the basis to determine appropriate thermal and flow measurement devices is defined. To properly choose the right solution, a short review of the available techniques and methods is also presented. Based on the given background, the design of the proposed test rig is presented, and final conclusions are given.

2. Materials and Methods

2.1. Conceptual Design of the Test Rig

The considered test rig is a part of the developed cycle of a negative CO_2 emission power plant using gasified sewage sludge as a main fuel—the assumptions and scheme of which were presented recently [11,12].

As the main theoretical assumptions of the whole system are known, the main aim of the part presented here was to develop a prototype research installation to study the performance of a spray-ejector condenser. Madejski et al. presented [32,33]; the study includes a description of the concept, schematics, operating ranges, proposals for acquisition and monitoring of operating parameters and basic assumptions for the implementation of the spray-ejector condenser installation. This applies in particular to the condensation of water vapour contained in a gas mixture (e.g., CO_2) and to the generation of a pressure rise in the gas mixture.

Based on the analysis of the sources presented in the introductory part, in order to build the test stand enabling the planned scope of research independently on other components, it was proposed to build a system with the necessary additional subsystems:

- Superheated steam generation system (I).
- CO_2 supply and mixing system (II).
- Spray ejector condenser with the water supplying system (III).
- Tank for water/steam/CO_2 mixture separation into gaseous and liquid parts with their common inlet and separate outlets (IV).
- Connection system between the spray ejector condenser and other parts (V).
- Measurement and monitoring system with data acquisition (VI).

A schematic diagram of the proposed test stand with all additional subsystems, in the form of a simulation model in Ebsilon software, is given in Figure 1. This model was then used to establish the minimum and maximum values of thermal and flow variables in the circuit necessary for the design of the test rig and the selection of the necessary measurement equipment. For the proposed nominal operating conditions of SEC, the thermal capacity of the proposed direct contact type condenser was around 22.5 kW and represents mainly the thermal energy of condensed steam.

The conceptual design of the prototype test stand included assumptions necessary to perform tests on a spray ejector condenser, especially the possibility of supplying the condenser with a liquefied medium (water vapour/CO_2 mixture) and with cooling water, discharging the mixture in the form of water/water vapour/CO_2 and separation of the gaseous part from the liquid (water) part. All tests will be supported by controlling and monitoring selected plant parameters.

The first subsystem is responsible for generating and supplying superheated steam with the possibility of temperature, pressure and steam mass flow control. It consists of a tank with distilled water (or a water treatment plant), a superheated steam generator with capacity regulation, an electric heater for additional steam superheating and necessary sensors of steam pressure, temperature and mass flow.

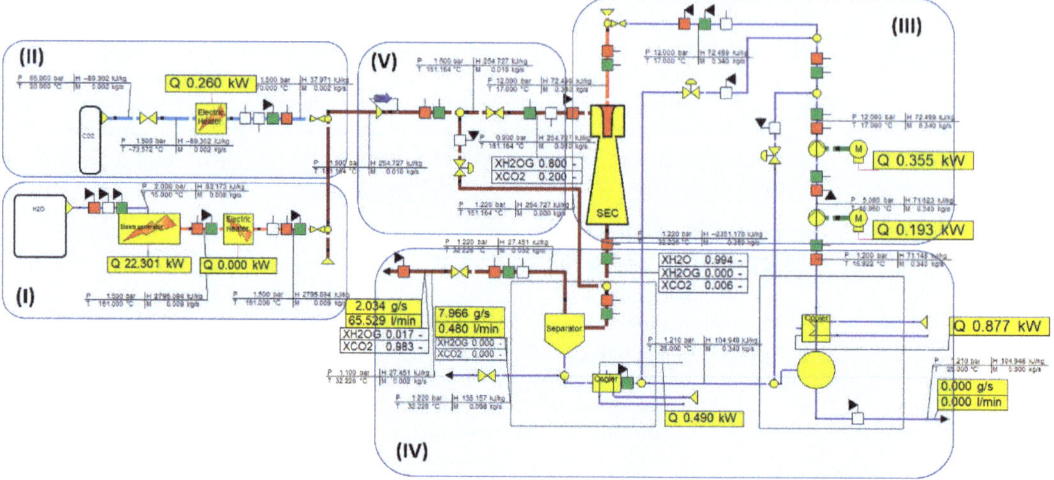

Figure 1. The concept of the designed test stand with the spray ejector condenser performances analysis opportunities. Superheated steam generation system (**I**), CO_2 supply and mixing system (**II**), spray-ejector condenser with the water supplying system (**III**), tank for water/steam/CO_2 mixture separation into gaseous and liquid parts with their common inlet and separate outlets (**IV**), connection system between the spray-ejector condenser and other parts (**V**).

The main task of the second subsystem is to deliver CO_2 with adjustable temperature, pressure and mass flow, together with the CO_2-steam mixing. It is built from a set of liquid CO_2 cylinders with the necessary equipment, an electric heater to warm up CO_2 after its expansion in order to prevent freezing, pressure, temperature and CO_2 mass flow sensors and mixing valves enabling the injection of CO_2 into the steam.

The third system is designed to supply the spray condenser with water at adjustable temperature, pressure and mass flow. It consists of a tank with water supplying a condenser and a cooler to decrease the water temperature, a supplying pump with controlled capacity, two bypasses to adjust water flow rate, and pressure, temperature and water flow rate sensors.

The next part enables separation of the water/steam/CO_2 mixture into gas and liquid parts along with the discharge of gases and liquids. Its main part is a mechanical separator to separate the gaseous part from the liquid part, the second tank in which the separator is immersed, together with the water-cooling system, gas outlet section from the tank with a valve to gas pressure control in the upper part of the separator, pressure and temperature sensors at the condenser's outlet and, additionally, for CO_2 mass flow at the outlet of the separator.

The main task of the fifth sub-system is to provide necessary connections for motive water (with possible length adjustment and change of inlet diameter), water vapour and CO_2 inlet and outlet with the water/CO_2 mixture pressure control.

The last part of the whole test rig is responsible for measurement of necessary temperatures, pressures and flow rates in selected points together with data visualisation and recording for further processing and analysis. In addition, control of the selected parameters for process fluids should be possible.

The selection of proper measurement devices is very important from the point of view of the planned experiments. On the other hand, the variety of available solutions is a difficult task. Hence, the following sections will be discussed in more detail, taking into account the specifics of the presented installation.

2.2. Mathematical Description of SEC

The main task of the designed test facility is to provide the base for the spray ejector condenser investigation. Hence, the mathematical description of this element is needed to identify the measured physical quantities necessary to evaluate the performance of the analysed process. Its schematic view is given in Figure 2.

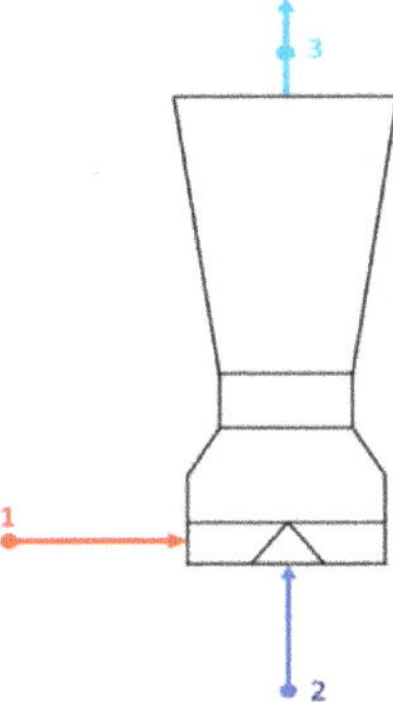

Figure 2. Scheme of spray-ejector condenser (1—suction fluid inlet, 2—motive fluid inlet, 3—outlet).

SEC has two inlets (for suction fluid and motive fluid) and one outlet. In case of water and steam SEC, according to energy balance, heat given by steam (suction fluid) is equal to heat absorbed by cooling water (motive fluid). From this, we obtain [34]:

$$\dot{Q}_s = \dot{Q}_{cw}, \tag{1}$$

and mass balance:

$$\dot{m}_s + \dot{m}_{cw} = \dot{m}_c. \tag{2}$$

When considering CO_2 added to suction fluid, then energy balance can be written in the following form:

$$\dot{Q}_{cw} = \dot{Q}_{vap\_C} + \dot{Q}_C + \dot{Q}_{cond\_c} + \dot{Q}_{CO2\_c}. \tag{3}$$

The cooling water draws heat from the cooling of the steam, cooling of CO_2, condensation of the steam and cooling of the condensed steam (water).

Heat of water vapour cooling $\dot{Q}_{vap\_C}$ is expressed as:

$$\dot{Q}_{vap\_C} = \dot{m}_{H_2O} \cdot \left(h_{1H_2O} - h_{sat\_vap}\right). \tag{4}$$

Heat of condensation Q_C was calculated according to the formula:

$$\dot{Q}_C = \dot{m}_{H_2O} \cdot r. \tag{5}$$

Heat of CO_2 cooling $\dot{Q}_{CO_2\_c}$ was calculated according to the formula

$$\dot{Q}_{CO_2\_c} = \dot{m}_{CO_2} \cdot \left(h_{1CO_2} - h_{3CO_2}\right). \tag{6}$$

Heat of water condensate cooling $\dot{Q}_{cond\_c}$ was calculated according to the formula:

$$\dot{Q}_{cond\_c} = \dot{m}_{H_2O} \cdot \left(h_{sat_{liq}} - h_{3H_2O}\right), \tag{7}$$

with:

$$h_{sat_{liq}} = h_{sat\_vap} - r. \tag{8}$$

Heat absorbed by cooling water can be expressed as:

$$\dot{Q}_{cw} = \dot{m}_{cw} C_w \cdot \Delta T. \tag{9}$$

Because the temperature of condensate is equal to cooling water temperature at the outlet, then:

$$\Delta T = T_c - T_{cw,in}. \tag{10}$$

The enthalpy of water vapour and liquid (h_{1H_2O}, h_{sat_vap}, r, $h_{sat_{liq}}$, h_{3H_2O}) can be obtained from IF-97 steam tables [35]. The enthalpy of carbon dioxide CO_2 (h_{1CO_2}, h_{3CO_2}) can be obtained from NIST tables [36].

As can be deduced from the above equations, for proper energy balance calculations, mass flow rate and temperature should be measured. Enthalpy is taken from various physical tables for given conditions defined by pressure and temperature. Hence, heat flows given in Equations (3)–(8) are measured indirectly.

When a given physical variable, y, is measured indirectly and is a function of independent measurements $x_1, x_2, \ldots, x_n$:

$$y = f(x_1, x_2, \ldots, x_n) \tag{11}$$

Using the propagation model of uncertainty [37–41], the standard combined uncertainty u_c of y can be calculated from the formula:

$$u_c(y) = \sqrt{\left(\frac{\partial y}{\partial x_1} u(x_1)\right)^2 + \left(\frac{\partial y}{\partial x_2} u(x_2)\right)^2 + \ldots + \left(\frac{\partial y}{\partial x_n} u(x_n)\right)^2}. \tag{12}$$

Then, the expanded uncertainty is calculated from the equation:

$$U = k\, u_c(y). \tag{13}$$

When considering Equation (4), it can be seen that the standard combined uncertainty can be written as:

$$u_c\left(\dot{Q}_{vap\_C}\right) = \sqrt{\left(\frac{\partial \dot{Q}_{vap\_C}}{\partial \dot{m}_{H_2O}} u(\dot{m}_{H_2O})\right)^2 + \left(\frac{\partial \dot{Q}_{vap\_C}}{\partial h_{1H_2O}} u(h_{1H_2O})\right)^2 + \left(\frac{\partial \dot{Q}_{vap\_C}}{\partial h_{sat\_vap}} u(h_{sat\_vap})\right)^2} \tag{14}$$

with partial derivatives, as follows:

$$\frac{\partial \dot{Q}_{vap\_C}}{\partial \dot{m}_{H_2O}} = h_{1H_2O} - h_{sat\_vap}, \tag{15}$$

$$\frac{\partial \dot{Q}_{vap\_C}}{\partial h_{1H_2O}} = \dot{m}_{H_2O}, \tag{16}$$

$$\frac{\partial \dot{Q}_{vap\_C}}{\partial h_{sat\_vap}} = \dot{m}_{H_2O}. \tag{17}$$

Uncertainty of mass flow rate measurement, $u(\dot{m}_{H_2O})$, can be easily estimated based on the manufacturer's data for a given device. The uncertainty concerning water enthalpy, and also saturated vapour, depends on estimation accuracy of this variable. According to IF-97 and NIST tables, within the region of interest, the enthalpy of steam is defined at a given temperature and pressure with the uncertainty of $\pm 0.2\%$ [42]. The specific enthalpy of CO_2 is provided with an uncertainty of 0.95% [43].

However, if indirect measurements are used, then enthalpy is estimated based on measured temperature and pressure. Then, the total uncertainty of its estimation should be calculated from the relationship:

$$u(h) = \sqrt{(u_{h,table})^2 + (u(p))^2 + (u(T))^2}, \qquad (18)$$

For the next element of heat balance, heat of condensation, given by Equation (5), it can be written:

$$u_c\left(\dot{Q}_C\right) = \sqrt{\left(\frac{\partial \dot{Q}_C}{\partial \dot{m}_{H_2O}} u\left(\dot{m}_{H_2O}\right)\right)^2 + \left(\frac{\partial \dot{Q}_v}{\partial r} u(r)\right)^2} \qquad (19)$$

and with the following partial derivatives:

$$\frac{\partial \dot{Q}_C}{\partial \dot{m}_{H_2O}} = r, \qquad (20)$$

$$\frac{\partial \dot{Q}_C}{\partial r} = \dot{m}_{H_2O}. \qquad (21)$$

The calculation of uncertainty for heat of CO_2 cooling, $\dot{Q}_{CO2\_c}$, and heat of water condensate cooling, $\dot{Q}_{cond\_c}$, results in equations in the same form as for heat of water vapour cooling, $\dot{Q}_{vap\_C}$. Hence, this procedure is not presented here.

For the next element, heat absorbed by cooling water, we obtain:

$$u_c\left(\dot{Q}_{cw}\right) = \sqrt{\left(\frac{\partial \dot{Q}_{cw}}{\partial \dot{m}_{cw}} u\left(\dot{m}_{H_2O}\right)\right)^2 + \left(\frac{\partial \dot{Q}_{cw}}{\partial C_w} u(C_w)\right)^2 + \left(\frac{\partial \dot{Q}_{cw}}{\partial \Delta T} u(\Delta T)\right)^2} \qquad (22)$$

with:

$$\frac{\partial \dot{Q}_{cw}}{\partial \dot{m}_{cw}} = C_w \cdot \Delta T, \qquad (23)$$

$$\frac{\partial \dot{Q}_{cw}}{\partial c_w} = \dot{m}_{cw} \cdot \Delta T, \qquad (24)$$

$$\frac{\partial \dot{Q}_{cw}}{\partial \Delta T} = \dot{m}_{cw} C_w. \qquad (25)$$

Taking into account Equation (9), it should be noted here that:

$$u_c(\Delta T) = \sqrt{\left(\frac{\partial \Delta T}{\partial T_c} u(T_c)\right)^2 + \left(\frac{\partial \Delta T}{\partial T_{cw,in}} u(T_{cw,in})\right)^2}, \qquad (26)$$

with:

$$\frac{\partial \Delta T}{\partial T_c} = 1, \qquad (27)$$

$$\frac{\partial \Delta T}{\partial T_{cw,in}} = -1. \qquad (28)$$

Using presented relationships, there can be estimated measurement uncertainties at given known operating conditions.

2.3. Design Assumptions

The input assumptions for the technical analysis and selection of measurement equipment were taken from the simulation model of the considered cycle (Figure 1). At first, rated values were assumed as follows: water inlet at 17 °C and 12 bar, gas inlet at 150 °C, 0.2/0.9 bar and 0.01 kg/s. Outlet pressure was established at 1.10/1.15 bar. These operating parameters are the result of the first stage of the condenser design. It should be then experimentally tested.Hence, at the first stage of simulations, it was established that the designed system has to allow for adjustment of the operating parameters in the range presented in Table 1.

Table 1. The range of operating variables.

Location	Temperature	Pressure	Mass Flow
	°C	bar	kg/s
Water inlet	5–20	2–16	Resulting
Gas inlet	110–170	0.2–2.0	0.008–0.015
Outlet	Resulting	1–3	Resulting

In the next step, detailed simulations were performed to establish operating conditions in both extreme (minimum/maximum) cases representing two variants of the condenser operation (gas pressure at the inlet 0.2 bar and 0.9 bar). Table 2 summarises the selected values in the most important points of the designed test rig.

Table 2. Simulated values of operating variables in two variants.

Variable	Variant 1	Variant 2
Steam/CO_2 mixture pressure at the inlet to SEC	0.9 bar	0.2 bar
Steam flow rate at the inlet of SEC	8 g/s	8 g/s
CO_2 flow rate at the inlet to SEC	2 g/s	2 g/s
Motive water flow rate at the inlet to SEC	340 g/s	4399 g/s
Mixture flow rate at the outlet of SEC	350 g/s	4409 g/s
Motive water temperature at the inlet to SEC	17.0 °C	17.0 °C
Average mixture temperature at the outlet of SEC	32.2 °C	18.4 °C
CO_2 temperature before mixing valve with steam	70.0 °C	70.0 °C
Steam temperature before mixing valve with CO_2	161.0 °C	161.0 °C
Steam/CO_2 mixture temperature at the inlet to SEC	151.2 °C	151.2 °C
Mass Entrainment Ratio	0.0294	0.0023
Volumetric Entrainment Ratio	56.38	19.62

The presented values set the operating conditions for further selection of the measurement equipment. For practical reasons,it was additionally assumed that the mass flow rate of steam, CO_2 and motive water can be controlled from zero to approximately 120% of the maximum values. This creates a safety margin for flow meters, especially those sensitive to exceeding rated operating conditions. Based on this, and taking into account the physical side of the considered phenomena, there were established borders defining the variability of basic thermodynamic parameters in the considered points (Table 3).

Table 3. Variability ranges of measured quantities in selected points.

Process Fluid	Temperature Range	Pressure Range	Mass Flow Rate Range	Desired Accuracy
	°C	bar	g/s	% mol
CO_2	20–70	1–1.5	0–2	–
Steam	100–170	1–1.5	0–8	–
CO_2 + steam	110–170	1–1.5	0–10	–
Water	10–20	10–12	0–6000	–
Water + CO_2 (dissolved)	15–30	1–1.5	0–6000	0.0.1–0.1

2.4. Selected Design Aspects

The design of the test rig depends on the selected specific solutions and technical requirements of manufacturers. Hence, at this stage, it is difficult to provide detailed considerations. For these reasons, the procedure to estimate pressure loss in a water supplying system (system III in Figure 1) is presented, which is crucial for selection of a pump.

We begin with calculations of hydraulic loss. The pipe diameter was set at DN 50 and, following Table 2, the two extreme values of the mass flow rate Q_{min} = 340 g/s and Q_{max} = 4399 g/s were assumed. The cross-sectional area for solution flow was for the diameter DN 50: A_{50} = 2.04 × 10^{-3}m^2. Hence, the minimum and maximum flow velocity are: w_{d50min} = 0.17 m/s and w_{d50max} = 2.16 m/s, respectively.

Then, the friction factor to obtain major (friction) losses in a pipe was determined. The Reynolds number was computed from the definitional relationship:

$$Re = \frac{w_r \cdot d}{\nu_r}.$$

(29)

The kinematic viscosity for water is ν_r = 9.79 × 10^{-7} m^2/s.
For the diameter DN 50, we obtain:

$$Re_{50min} = \frac{0.17 \times 0.051}{0.000000979} = 8856,$$

(30)

$$Re_{50max} = \frac{2.16 \times 0.051}{0.000000979} = 112523.$$

(31)

The friction factor of major losses, λ, was calculated from the Nikuradse formula:

$$\frac{1}{\sqrt{\lambda}} = 2\log\left(\frac{D}{k} + 1.14\right).$$

(32)

where:

k—average roughness of the pipeline; for new pipes k = 0.02–0.10 mm.

It was assumed that k = 0.008 mm. From the above, for the diameter DN = 50 mm, the average friction factor λ = 0.241.

The pressure loss in a straight pipe section is given by the Darcy–Weisbach formula:

$$\Delta p_l = \lambda_{50}\frac{l}{d_{50}} \cdot \frac{\rho w^2}{2}.$$

(33)

The pressure loss per unit length of a straight pipe of 1 m length is:

$$\Delta p_{50/m} = \lambda_{50}\frac{1}{d_{50}} \cdot \frac{\rho w^2}{2}.$$

(34)

From the above: $\Delta p_{50/m\ max}$ = 11010 Pa/m and $\Delta p_{50/m\ min}$ = 68 Pa/m.

Based on the initial design, the total length of the straight pipe sections was assumed to be L = 12 m. From Equation (37) and for DN = 50 mm, we obtain Δp_{50Lmin} = 816.0 Pa and Δp_{50Lmax} = 132.12 kPa.

Minor pressure losses are a consequence of the presence of elements disturbing the flow in the pipeline-elbows, fittings, etc. Accurate determination of the value of these losses requires precise knowledge of the flow characteristics of these elements.

When changing the flow direction, the local resistance coefficient can be determined from the relationship:

$$\xi = 0.946\sin^2\left(\frac{90}{2}\right) + 2.05\sin^4\left(\frac{90}{2}\right).$$

(35)

When changing the flow direction by $90°$, the local resistance coefficient has the value $\xi = 0.9855$, and the minor pressure loss is given by the relationship:

$$\Delta p_m = \xi \frac{\rho w^2}{2}. \tag{36}$$

From Equation (40) for the DN50 diameter, we obtain: $\Delta p_{50/\xi,\,min} = 14$ Pa, $\Delta p_{50/\xi,\,max} = 2299$ Pa. The change of direction at an angle of $90°$ in the system will take place $n = 3$ times, then the total minor losses. Hence, $\Delta p_{\xi 50\,min} = 42$ Pa and $\Delta p_{\xi 50\,max} = 6.897$ kPa.

The remaining local resistances resulting from the installed fittings and instrumentation were replaced with the equivalent linear resistances, assuming the equivalent length $L_Z = 5$ m. For the equivalent length, we determined the values of minor pressure losses. For the DN50 diameter, we obtain:

$$\Delta p_{50Z} = \Delta p_{50/m} \cdot L_Z. \tag{37}$$

From Equation (40): $\Delta p_{50zmin} = 68$ Pa/m $\times$ 5 m $= 340$ Pa and $\Delta p_{50zmax} = 11.01$ kPa/m $\times$ 5 m $= 55.05$ kPa.

The total minor losses in the system are the sum of the bend losses and the equivalent minor losses:

$$\Delta p_{50m} = \Delta p_{50Z} + \Delta p_{\xi 50}. \tag{38}$$

From the above, we obtain $\Delta p_{50mmin} = 340 + 42 = 382$ Pa and $\Delta p_{50mmax} = 55.05 + 6.897 = 61.95$ kPa. Total pressure in the pipeline is the sum of minor and major pressure losses:

$$\Delta p_r = \Delta p_L + \Delta p_m. \tag{39}$$

Its minimum and maximum values are then $\Delta p_{50rmin} = 816 + 382 = 1.2k$ Pa and $\Delta p_{50rmax} = 132.12 + 61.95 = 194.07$ kPa, respectively. Hence, the total maximum pressure loss was assumed to be 194.1 kPa, and this value was used to select the water pump.

3. Selection of Measurement Techniques and Devices

3.1. Temperature

Temperature measurement is very important in numerous technical, industrial and scientific applications. The accuracy of a performed measurement depends on technical and physical factors and relies on the proper selection of the measuring method and device. This is influenced by the kind of medium and the dynamics of physical phenomena occurring within the studied area. They can be defined before the measurements are carried out. Depending on the amount of input design data, there can be defined measurement conditions necessary for the choice of the appropriate measurement method and tools. Due to the use of heat transfer mechanisms, two kinds of temperature measurement methods are defined, i.e., contacting or non-contacting methods [44].

As far as the designed test bed is concerned, the temperature range of working mediums is from 10 °C to 180 °C. Hence, both contact and non-contact methods can be applied here. For practical reasons, contact methods applying thermocouples (TC) and resistance temperature detectors (RTD) are the most useful. This is because both these kinds of sensors provide electrical-type output measurement signals, which can be very easily transmitted and converted in computer measurement systems [45].

Depending on the kind of the resistive material used, currently, RTDs are manufactured in three variants, namely, as copper, nickel or platinum resistors. The use of the first two is typically restricted to the range from −50 °C to +160 °C. The most popular are platinum sensors due to their good accuracy, low cost and reproducibility. Their application covers the range from −200 °C to +850 °C. Thin-film resistors can be very small with the length of 3 mm. They also offer high accuracy. Following the IEC 60751 standard [46], at a temperature of 200 °C, the wire-wound A-class sensor has the tolerance of ±0.44 °C. However, the platinum resistive wire is completely encapsulated to avoid a negative external environment (mechanical stress, corrosion, oxidation, etc.). However, it results in poorer long-term stability and large hysteresis due to differences in thermal expansion of the substrate and the wire and the indirect contact of the wire and process medium [47]. This effect is enhanced when placing the encapsulated sensor in a protective sheath.

The operation of thermocouples is based on Seebeck's thermoelectric effect. They offer significantly wider measurement range, from $-270\,^\circ$C to $+1370\,^\circ$C (K-type nickel–chromium alloy thermocouple) to over 2000 $^\circ$C (Pt-Rh thermocouples). Tolerance classes are given in IEC 60584 [48], and for class 1 of a sensor at a temperature of 200 $^\circ$C,the tolerance is from $\pm0.50\,^\circ$C (type T) to $\pm1.50\,^\circ$C (types E, J, K and N).

The sensitive measuring point of the thermocouple is the measuring junction, where the electrical and thermal contact between two conductors. As this junction can be very small (diameter below 0.5 mm), thermocouples may offer low short response time and are better suited to measurements when thermal dynamics are of greater importance. This type of junction is called exposed. However, to protect against the negative impact of the external environment, a junction is mounted in a suitable protective tube, also named a well or sheath. In such a case, a junction can be mounted in two variants. The first one is ungrounded, a junction suspended inside the metallic sheath, such that the junction is not in contact with the sheath. It causes the thermocouple to react to the temperature change in the environment longer because the sheath must reach thermal equilibrium before the junction can detect that change. In the second one, with a grounded junction, a junction is welded to the inside of the sheath resulting in a shorter reaction time to temperature changes.

To clarify this issue, in Table 4, on the base of manufacturers catalogues [49], the time after which the sensor's response to a step change in temperature reaches 90% of its maximum value in case of isolated and grounded thermocouples in moving water (0.4 m/s) and air (2 m/s)is given.For comparison, the same parameter was also given for RTD sensors of the same manufacturer in the same flow conditions in Table 5. It is clearly visible that they have a longer response time. Hence, this parameter should be balanced with accuracy when considering applications requiring more precise dynamic process analysis.

Table 4. The time after which the thermocouple's response to a step change in temperature reaches 90% of its maximum value, s.

Sheath Diameter/Junction Type	Water 0.4 m/s	Air 1 m/s
1 mm/grounded	0.18	10
1 mm/insulated	0.50	10
3 mm/grounded	0.75	80
3 mm/insulated	2.90	88
6 mm grounded	2.60	185
6 mm/insulated	9.60	200
8 mm/grounded	3.90	250
8 mm/insulated	14.00	290

Table 5. The time after which the RTD's response to a step change in temperature reaches 90% of its maximum value, s. Steel sheath.

External Diameter and Wall Thickness, mm	Water 0.4 m/s	Air 1 m/s
6×0.5	55	260
10×1.5	100	400
15×1.5	170	490
22×2.0	480	1200

The maximum process fluid velocity is 8.9 m/s, 18.5 m/s, 24.8 m/s and 3.3 m/s for CO_2, steam, steam + CO_2 and water, respectively. These values guarantee that the thermal time constant should decrease by over 3–4 times in relation to still air or water [50].

The most important advantages of RTDs are: good accuracy, stability, sensitivity and reproducibility. On the other hand, they are more expensive, larger and more fragile than thermocouples.

It should also be noted that when an RTD insulation resistance is affected by moisture, an additional error is introduced. Hence, Pt100 sensors offer better long-term stability when compared to Pt1000. At the same time, the resistance of lead wires should be taken into account, and for this reason, 4-wire and 3-wire measurement circuits are preferred.

Thermocouples are relatively inexpensive, manufactured in a variety of sizes and cover a wide temperature range. They are smaller than RTDs, reasonably stable, reproducible and fast. Additionally,thermoelectric force is independent of wire length and diameter.

The low output signal from TCs is sensitive to electrical noise and requires additional amplifiers. Furthermore, bare thermocouples cannot be used in conductive fluids, and the thermoelectric junction is sensitive to stress and vibration.

In the designed test rig,the ambient temperature should not exceed the typical values of 20–30 °C met in indoor conditions. Therefore, its impact on thermal conditions of the rig can be omitted. Research on SEC operation is planned to be conducted in thermally established conditions. Gas, steam and water flows in piping are supposed to be well developed. The distance between the data logger and measuring points (length of lead wires) should not exceed 3 m. Hence, A-class Pt100 sensors were chosen for temperature measurements.

3.2. Pressure

Piezoelectric transducers are commonly used among commercially available electric pressure transducers due to their accuracy, stability, linearity and favourable price. To reduce temperature errors, the elastic element (membrane) should be protected against the action of the high-temperature fluid. Hence, in the considered test rig selection, pressure sensors should be based not only on the expected pressure values but on the maximum temperature of process fluid.

In case of process temperatures over 200 °C,a non-insulated siphon tube is usually installed between the process connection and the pressure sensor. In that tube, condensate or chilled water accumulates, protecting the elastic element against high fluid temperature. Its length depends on the process temperature and maximum working temperature of the pressure sensor.

It may be sufficient to use the special finned cooling element in applications for lower process temperatures, usually with 3 or 5 fins.

3.3. Flow Rate

As given in Section 2, four measuring points were selected: CO_2, steam, steam + CO_2 and motive water. A two-phase flow is not meant to occur. Only in one case will a mixture of gases flow. Temperature and pressure conditions are given in Table 2.

Mass flow rate measurement is more convenient than volume flow from the point of view of energy balance equations in thermodynamic systems. However, both mass and volume flow measurement methods and devices are considered in this paragraph because all of them have certain advantages that can be decisive when choosing a given solution.

The operation of differential pressure flow meters (orifice plate, Venturi tube, and nozzle) is based on Bernoulli's equation. They are simple and reliable devices when manufactured and installed following relevant standards. Their main disadvantage is pressure drop and their sensitivity to installation conditions. They are commonly used in fluids and dry gases mass flow measurements. In case of two-phase flows, a broad range of empirical models were developed to correlate mass flow rate with pressure drop for different types of flow and fluid combinations.

Table 6 presents selected results for the design of orifice flow meters for the considered four points following ISO/TR 15377:2007 and EN ISO 5167:2005 standards. In all cases, a single 90° elbow or tee (single branch flow) at the inlet of the meter was assumed. Pressure drop is given at maximum mass flow.

Table 6. Selected parameters of orifice plates.

Process Fluid	Inlet Length	Outlet Length	Minimum Flow	Maximum Flow	Pressure Loss
	m	M	kg/s	kg/s	kPa
CO_2	0.56	0.22	0.000	0.003	0.135
Steam	1.20	0.20	0.002	0.015	1.861
CO_2 + steam	1.20	0.20	0.002	0.015	1.861
Water	2.40	0.43	0.219	6.000	9.050

Among variable area flow meters, rotameters are very popular because of their simple and reliable construction, low pressure drop and applicability to various clean gases and liquids, because sediments and gas bubbles affect accuracy. For gases, the effect of pressure and temperature is taken into account using scale correction factors. Typical accuracy is 1 to 4% of the full scale.

The main part of a turbine flow meter is a turbine rotor. It is concentrically mounted on a shaft in a pipe with a measured flow Rotation velocity rate depends on fluid volume passing in a given time step. The typical accuracy is 0.25%.

The operation of the electromagnetic flow meters is based on Faraday's law of electromagnetic induction. When the conductive liquid passes through a channel or a pipe, an electromotive force (voltage) is induced and measured by two electrodes mounted flush to the pipe wall. This voltage depends proportionally on the average flow velocity (and volume flow rate). Because the measured medium must be conductive, that type of flow meter is not suitable for gas flow. Typical accuracy can be 0.5% to 1%, but it can be worsened by impurities in the liquids.

Ultrasonic flow meters can be applied in a broad range of fluids providing no obstruction to flow. Two types are in use: Doppler and transit time.

In the Doppler flow meter, pulses of high-frequency sound waves are sent across the pipe and then received by the detector. The wave moving with the flow travels faster than that moving against the flow. The time difference between these two periods depends on the fluid flow rate. Hence, this kind of ultrasonic flow meter requires clean fluids for proper operation because sound waves are reflected or scattered from suspensions in the flow path.

The transit time flow meter measures the time difference between two ultrasonic pulses transmitted against and with the flow. This device is also sensitive to entrained gas bubbles and solid particles.

The operation of vortex flow meters is based on the Karman vortex streets phenomenon. The main advantage of this meter is the independence of measured volumetric flow on the fluid density, simple construction, lack of moving parts and good operation with liquids and gases, but it creates a pressure drop comparable to the orifice plate or turbine meters. Typically, it has 1% accuracy.

Coriolis flow meters directly measure the mass flow of fluids of different densities and viscosities. They can also measure the mass flow of two-phase mixtures, liquid–liquid (such as emulsions) and liquid–solid [51].

Their principle of operation is based on the use of the Coriolis phenomenon, which causes the occurrence of inertial forces in the case of simultaneous rotational and progressive motion of the body. The direction of the inertia force is determined by the vector product of the linear velocity and the rotational speed. In flow meters using this phenomenon, the rotational motion was replaced by the vibrating motion of the measuring tubes. The measure of the mass flux is the value of the phase shift of vibrations at the inlet and outlet of these tubes (often two are used).

Thermal flow meters utilise the principle that the rate of heat absorbed by a flowing fluid is proportional to its mass flow rate.

Two types of thermal mass meters are in use: constant temperature and constant current. In the first case, the temperature rise (which depends on the mass flow) of the flowing fluid is measured at a fixed heat transfer rate. In the second case for a fixed temperature rise, the heat transfer rate is measured. Its value increases with the fluid mass flowrate. These devices can be used for liquid flows, but practically,they are used mainly for gas flow measurements. Their typical accuracy is 2%.

In order to choose reliable devices among commercially available devices while considering economical and technical constraints, a more detailed analysis was performed. First, questionnaires were sent with operational conditions for selected measuring points to the several leading manufacturers. The obtained answers formed the basis for further considerations.

As given in Table 7, the most economically competitive solution is an orifice flow meter. They are simple and reliable, but additional measurements of fluid pressure and temperature should be performed to measure mass flow. Coriolis mass flow meters provide a reliable measurement but are the most expensive. Hence, other factors were taken into account.

Table 7. Types and prices (EUR net, 04.2022) of flow meters proposed by different suppliers.

Process Fluid	1	2	3	4
CO_2	Orifice/990	Thermal [1]/3900	Coriolis [1]/4900	Rotameter/1500
Steam	Orifice/835	Vortex [1]/3000	Coriolis [1]/4000	Vortex/2500
CO_2 + steam	Orifice/835	Vortex/2100	Coriolis [1]/4000	Vortex/2500
Water	Orifice/875	Electromagnetic/1750	Coriolis [1]/9800	Coriolis [1]/5800

[1] Mass flow measurement.

Table 8 shows the declared accuracy of all devices. In all cases, the presented values include the total accuracy including the error of the volume/mass flow transducer and the electronic converter. The Coriolis flow meters offer the best accuracy in mass flow measurements but are also expensive. An interesting option for CO_2 is the thermal mass flow meter.

Table 8. Maximum errors of studied solutions at rated flow, %.

Process Fluid	1	2	3	4
CO_2	0.85	1.00	0.50	1.60
Steam	1.32	1.10	0.10	1.50
CO_2 + steam	1.32	1.10	0.10	1.50
Water	0.92	0.53	0.10	0.15

The next parameter to be considered is pressure loss. Its value affects flow conditions and must be considered when choosing a circulating pump or the place where pressure transducers will be installed. Its maximum values are given in Figure 3. The fourth supplier did not provide the information about pressure loss for water flow measurement. In the case of the second manufacturer, the proposed electromagnetic device has no pressure loss.

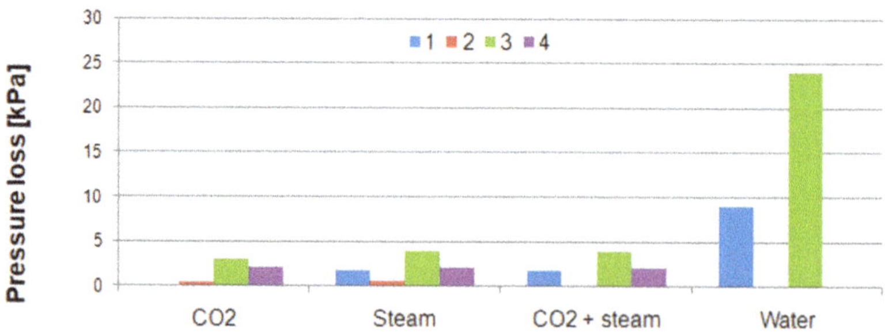

Figure 3. Maximum pressure loss at rated flow.

Table 9 shows selected solutions. The second solution was evaluated as the most competitive. The thermal flow meter for CO_2 was selected because of this reasonable price and good quality. For steam measurement, the vortex mass flow meter was chosen. In the final case, motive water, the electromagnetic device was selected. It measures volume flow. However, as motive water temperature is supposed to vary within a narrow range, the mass flow rate can be easily and precisely measured from volume flow rate and water temperature. It is also very important that this type of flow meter has negligible pressure loss in relation to other solutions: 24 kPa for the Coriolis flow meter and 9.05 kPa for the orifice meter, which significantly impacts pump selection and operation.

Table 9. Selected parameters of chosen solutions.

Process Fluid	Type	Minimum Flow	Maximum Flow	Pressure Loss
		kg/s	kg/s	kPa
CO_2	Thermal	0.0005	0.002	0.05
Steam	Vortex	0.005	0.013	0.67
CO_2 + steam	Vortex	0.005	0.010	0.40
Water	Electromagnetic	0.100	6.000	0.00

3.4. CO_2 Content

The measurement of CO_2 content, both dissolved in water and gaseous in air, is technically demanding due to this gas's physical and chemical properties. Several methods are currently in use in technical applications.

On the base of the conducted literature review [52–54] several CO_2 sensors were selected that possibly would be applicable in the designed test rig. They are as follows:

- Potentiometric sensors—Severinghaus,
- Potentiometric sensors—Solid electrolyte sensors,
- Non-dispersive infrared (NDIR) CO_2 sensors.

The main disadvantage of the potentiometric sensor is the necessary calibration procedure before measurements. It is difficult if measurements must be carried out under field conditions or for solutions with unknown and changing composition. Their typical response time is between 30 s and 120 s. The operation temperature is up to 50 °C.

Solid electrolyte sensors are, in principle, similar to the abovementioned. The operational temperature is usually within the 200–750 °C range. For semiconductor sensors, it is from −10 to 100 °C. Additional external power is required to heat up the sensor. The output of the electrical signal depends on gas concentration. Typical response time is approximately 1 s. These sensors do not require calibration but can be used only in gas applications.

The operation of the non-dispersive infrared (NDIR) sensors is based on the concentration-dependent absorption of electromagnetic radiation in the IR range. Their response time is usually between 5–30 s. Modern solutions are equipped with in-built converters and self-calibration functions. A short comparison of these sensors is given in Table 10.

Table 10. Selected parameters of CO_2 concentration measurement sensors.

Type	Response Time T0.9	Operating Temperature	Calibration	External Power
	s	°C	-	-
Severinghaus	30–120	<50	Yes	No
Solid electrolyte	<1	30–120	No	Yes
NDIR	5–30	0–50	No	Yes

On the basis of the presented review, recommended solutions were chosen. Three possible measurement points in the designed test rig were selected:

- CO_2 dissolved in outlet water from SEC,

- gaseous CO_2 at the outlet from the separator,
- CO_2 inside SEC.

For the first point, the Severinghaus-type potentiometric sensor can be recommended. Despite a troublesome initial calibration procedure, it can be used, as there are assumed stable operating conditions. However, the minimum response time should be thoroughly checked at first. The concentration of gaseous CO_2 can be easily determined using an NDIR sensor [53]. In the last case, inside SEC, a currently commercially available solution at a reasonable price cannot be proposed.

3.5. Data Acquisition

As the data acquisition system, a device with simultaneous sampling of measured inputs should be used. The possible solution should allow data recording with time stamps from approximately 0.1–1 s to 10 min.

The type of analogue measurement signals is of secondary importance. Typical output signals from sensors are recommended, both in voltage (0–10 V) or in current (4–20 mA) standards. The latter is safer from a practical perspective, because zero current means failure occurred in the circuit. In the case of voltage standard, this state can be interpreted as zero value of the measured quantity or as a fault in the system.

Temperature sensors require additional temperature converters. Pressure transducers have in-built converters. Flow meters are usually equipped with current (4–20 mA) and pulse/frequency outputs proportional to the value of the measured quantity.

4. Results and Discussion

Following the design assumptions presented in previous sections, input parameters for uncertainty calculations were chosen. They are given in Figure 1 and Table 2 as Variant 1. Based on these parameters, the thermodynamic parameters of water, steam and CO_2 in characteristic points of a spray ejector condenser are given in Table 11.

Table 11. Thermodynamic parameters of water, steam and CO_2 in characteristic points of a spray ejector condenser.

Variable	1	2	3
$T\ [\degree C]$	151.16	17.0	32.23
$p\ [bar]$	1.50	12.0	1.22
$h_{H2O}\ [kJ/kg]$	2775.25	72.50	135.16
$h_{CO2}\ [kJ/kg]$	620.18	-	511.83

In addition, from NIST tables, $p = 1.50$ bar: $h_{satvap} = 2693.13$ kJ/kg, $h_{satliq} = 467.0807$ kJ/kg and $r = 2226.033$ kJ/kg were estimated.

As the full uncertainty analysis, presented in Section 2.3, is too complicated to be presented here for all possible variants and variables, we presented an analysis for heat flux from water vapour cooling.

Taking values from Table 11, Equations (15)–(17) provide results as follows:

$$\frac{\partial \dot{Q}_{vap\_C}}{\partial \dot{m}_{H_2O}} = 2775.25 - 2693.13 = 82.19\frac{kJ}{kg}, \tag{40}$$

$$\frac{\partial \dot{Q}_{vap\_C}}{\partial h_{1H_2O}} = 0.008\frac{kg}{s}, \tag{41}$$

$$\frac{\partial \dot{Q}_{vap\_C}}{\partial h_{sat\_vap}} = 0.008\ \frac{kg}{s}. \tag{42}$$

The uncertainty of the mass flow measurement, $u(\dot{m}_{H_2O})$, can be taken directly from the manufacturer's datasheet and depends on the chosen device.

Taking the uncertainty of the steam enthalpy estimation as 0.2% (see Section 2.2), we obtain:

$$u(h_{1H_2O}) = 2775.25 \times \frac{0.2}{100} = 5.55\frac{kJ}{kg}. \tag{43}$$

Taking the uncertainty of the steam enthalpy estimation as 0.2% (see Section 2.2), the accuracy of pressure measurement as 0.5% and temperature as 0.1% for h_{sat_vap} = 2693.13 kJ/kg, we obtain from Equation (18):

$$u(h_{sat\_vap}) = \sqrt{(u_{h,table})^2 + (u(p))^2 + (u(T))^2} = 14.75\frac{kJ}{kg}. \tag{44}$$

Assuming the uncertainty of mass flow measurement $u(\dot{m}_{H_2O}) = 0.1\%$ and substituting the obtained values in Equation (19), we obtain:

$$\frac{\partial \dot{Q}_{vap\_C}}{\partial \dot{m}_{H_2O}}u(\dot{m}_{H_2O}) = 8.219 \text{ kg/s}, \tag{45}$$

$$\frac{\partial \dot{Q}_{vap\_C}}{\partial h_{1H_2O}}u(h_{1H_2O}) = 44.404 \text{ kg/s}, \tag{46}$$

$$\frac{\partial \dot{Q}_{vap\_C}}{\partial h_{sat\_vap}}u(h_{sat\_vap}) = 118.006 \text{ kg/s}, \tag{47}$$

and:

$$u_c(\dot{Q}_{vap\_C}) = \sqrt{(8.219)^2 + (44.404)^2 + (118.006)^2} = 126.35 \text{ W} \tag{48}$$

In the same way, $u_c(\dot{Q}_{vap\_C})$ from Equation (19) can be computed. The results for the various considered flow meters were as follows: 126.33 W, 126.31 W, 126.09 W and 126.77 W for orifice, first vortex, Coriolis and second vortex flow meters, respectively.

Several authors presented their concepts of test rigs with direct contact heat exchangers. The most important findings related to measurement devices are presented in the following paragraphs.

In test rigs with direct contact condensers for the temperature measurement of process fluids, both thermocouples and resistance sensors were used. In the first group, various types were chosen. The most popular was the K-type (nickel–chromium/nickel–alumel) thermocouple with a typical uncertainty of ±1.5 °C, within the range from −40 °C to 1000 °C [55,56]. However, sometimes authors used specially calibrated thermocouples to obtain better accuracy, as of ±1 °C [57,58], ±0.2 K in the range of 0–600 °C [59], 0.5% within the range 0 °C–200 °C [60] or of 0.1% [61] at the same conditions.

The latter group of sensors was represented mainly by Pt100 sensors [62] with typical accuracy of ±0.1 °C [63–65].

However, authors did not provide any further details on selection procedure to choose the given type of sensor in certain application.

For pressure measurement, mainly piezoresistive absolute and differential pressure transducers were used, with an uncertainty of 2% at temperatures up to 50 °C or 4% up to 80 °C [55,56], ±0.5% of full-scale (FS) with a measurement range of 0–25 MPa [66] or of 0.1% FS and range from 200 to 700 kPa [67]. The accuracy of the sensors in high temperature steam measurements was reported as 0.1% (0–1 MPa) [60] or 0.2% (0–1.6 MPa) [59].

When analysing published studies, the flow measurements of water and steam involved the application of a large variety of techniques, from classical methods to modern electronic devices. Genić et al. [63] applied orifice flow meters (meeting ISO 5167-1 requirements) with mercury U-tube manometers. In the case of water flow measurement, however, electromagnetic and turbine flow meters prevailed (Table 12). When steam flow rate is considered, vortex flow meters were the most popular (Table 13). However, despite a wide range of analysed studies, no selection guidelines were given by the authors.

Table 12. Devices used in water flow measurements.

Type	Range	Error	Reference
Rotameter	1–7 m^3/h	0.1 m^3/h	[68]
Electromagnetic	0.08–2.78 kg/s	0.2%	[60]
Electromagnetic	0.88–17.66 m^3/h	0.5%	[59]
Electromagnetic	0–10 m^3/h	1.0%	[66]
Rotameter	-	1.25%	[58]
Electromagnetic	0–30 m^3/h	0.5%	[28]
Turbine	0.04–0.25 m^3/h	1.0%	[59]
Turbine	0.6–6 m^3/h	1.0%	[59]
Rotameter	1–10 m^3/h	1.5%	[59]
Turbine	0.9–13.6 m^3/h	0.15%	[59]

Table 13. Devices used in steam flow measurements.

Type	Range	Error	Reference
Orifice	-	1.0 Pa	[68]
Vortex	7.5–73.9 g/s	1.0%	[60]
Vortex	30 to 300 m^3/h	1.5%	[59]
Vortex	0–40 m^3/h	0.75%	[28]
Vortex	0–120 m^3/h	0.75%	[28]
Vortex	0–150 kg/h	1.0%	[59]
Vortex	-	2.0%	[64]
Orifice	-	1.5%	[64]

Some authors also applied flow visualisation techniques. For example, Zong et al. [69] presented an experimental study on the direct contact condensation of steam jet in subcooled water flow in a rectangular mix chamber.

The steam and water temperature were measured by the K-type thermocouples (diameter of 1 mm, measurement range of 0–200 °C) with a maximum error of 0.5%. The 0–1 MPa, high-temperature pressure transducers with maximum error of 0.1% were used. A vortex flow meter with the range of 2700–26,600 kg/h and a maximum error of 1.0% was used for steam flow rate measurements. Additionally, a 330 kW electric steam generator with a maximum capacity of 0.11 kg/s was used as a steam source. The steam flow rate was controlled manually by a valve. The water flow rate was measured by an electromagnetic flow meter of 290–10,000 kg/h with a maximum error of 0.2%. The water flow rate was controlled by valves on feeding and bypass lines.

To record and visualise the flow field of the direct contact condenser, the Phantom V611 type high-speed camera, with 5 kHz frequency, was used. The LabVIEW data acquisition system, including an industrial computer, a National Instruments cDAQ-9178 data acquisition module, two NI 9213 modules (for temperature sensors)and four NI 9203 modules (for pressure transducers)were used.

The presented cases reveal a variety of used measuring devices. Thermocouples were usually used in temperature measurements due to their low thermal time constant. The piezoelectric transducer is a common solution in pressure measurement. Flow rate is the most problematic, and it should be noted here that applied solutions depend not only on the technical requirements but also on economic factors.

5. Conclusions

The presented concept of the laboratory test rig to experimentally investigate an application of a spray condenser in a CO_2 capture system in a thermal power plant cycle is novel and requires an in-depth analysis of the available sources in terms of the proper selection of measurement equipment.

The indicated solutions were chosen based on detailed simulations in a well-established environment. The provided results were then used as boundary conditions to choose the measurement equipment.

This part of the design procedure is very problematic as various technical and economic factors should be taken into account. However, comparison with other works showed that the manual control of process parameters is sufficient during operation of the studied cases, and stable thermal conditions were easily achieved. For these reasons, the dynamic phenomenon can be treated as less important, and more attention can be given to the analysis of the steady states, which are the normal operating states of thermal power plants.

Author Contributions: Conceptualization, P.M. (Paweł Madejski) and K.B.; methodology, P.M. (Paweł Madejski), P.M. (Piotr Michalak), M.K. and K.B.; software, P.M. (Paweł Madejski); validation, M.K., P.M. (Piotr Michalak) and T.K.; formal analysis, P.M. (Paweł Madejski) and P.M. (Piotr Michalak); investigation, P.M. (Paweł Madejski), P.M. (Piotr Michalak) and M.K.; resources, P.M. (Paweł Madejski) and P.M. (Piotr Michalak); data curation, P.M. (Paweł Madejski); writing—original draft preparation, P.M. (Piotr Michalak) and P.M. (Paweł Madejski); writing—review and editing, P.M. (Piotr Michalak) and P.M. (Paweł Madejski); visualization, P.M. (Piotr Michalak) and P.M. (Paweł Madejski); supervision, P.M. (Paweł Madejski); project administration, P.M. (Paweł Madejski); funding acquisition, P.M. (Paweł Madejski) and K.B. All authors have read and agreed to the published version of the manuscript.

Funding: The research leading to these results received funding from the Norway Grants 2014-2021 via the National Centre for Research and Development. The work was prepared within the frame of the project: "Negative CO_2 emission gas power plant"-NOR/POLNORCCS/NEGATIVE-CO2-PP/0009/2019-00, which is co-financed by the programme "Applied research" under the Norwegian Financial Mechanisms 2014–2021 POLNOR CCS 2019-Development of CO2 capture solutions integrated in power and industry processes.

Data Availability Statement: Not applicable.

Conflicts of Interest: The authors declare no conflict of interest.

Nomenclature

$\dot{Q}_C$	heat flux from steam condensation, W
$\dot{Q}_{cond\_c}$	heat flux from water condensate cooling, W
$\dot{Q}_{CO2}$	heat flux from CO_2 condensation, W
$\dot{Q}_s$	heat flux given by the steam, W
$\dot{Q}_{cw}$	heat flux absorbed by the cooling water, W,
$\dot{Q}_{vap\_C}$	heat flux from water vapour cooling, W
T_c	condensate temperature, K
$T_{cw,in}$	temperature of inlet cooling water, K
ΔT	cooling water temperature difference, K
c_w	specific heat of water, J/(kg·K)
d	internal pipe diameter, m,
h_{1CO_2}	carbon dioxide enthalpy at point 1, kJ/kg
h_{3CO_2}	carbon dioxide enthalpy at point 3, J/kg
h_{1H_2O}	water vapour enthalpy at point 1, J/kg
h_{3H_2O}	water vapour enthalpy at point 3, J/kg
h_{sat_liq}	saturated liquid enthalpy, J/kg
h_{sat_vap}	saturated vapour enthalpy, J/kg
K	coverage factor; k = 2 for 95% confidence level of the uncertainty,
$\dot{m}_c$	mass flow rate of the condensate, kg/s

$\dot{m}_{cw}$	mass flow rate of cooling water, kg/s
$\dot{m}_{s}$	mass flow rate of exhaust steam, kg/s
$\dot{m}_{H_2O}$	water vapour mass flow rate, kg/s
$\dot{m}_{cw}$	cooling water mass flow rate, kg/s
$\dot{m}_{CO_2}$	carbon dioxide mass flow rate, kg/s
u	uncertainty,
U	expanded uncertainty,
r	evaporation energy at saturation pressure, J/kg
$u_{h,table}$	uncertainty of specific enthalpy given in tables, J/kg
$w_{h,table}$	kinematic viscosity, m^2/s
w_r	water velocity, m/s

References

1. Ziembicki, P.; Kozioł, J.; Bernasiński, J.; Nowogoński, I. Innovative System for Heat Recovery and Combustion Gas Cleaning. *Energies* **2019**, *12*, 4255. [CrossRef]
2. Tic, W.J.; Guziałowska-Tic, J. The Cost-Efficiency Analysis of a System for Improving Fine-Coal Combustion Efficiency of Power Plant Boilers. *Energies* **2021**, *14*, 4295. [CrossRef]
3. Damartzis, T.; Asimakopoulou, A.; Koutsonikolas, D.; Skevis, G.; Georgopoulou, C.; Dimopoulos, G.; Nikolopoulos, L.; Bougiouris, K.; Richter, H.; Lubenau, U.; et al. Solvents for Membrane-Based Post-Combustion CO_2 Capture for Potential Application in the Marine Environment. *Appl. Sci.* **2022**, *12*, 6100. [CrossRef]
4. Slavu, N.; Badea, A.; Dinca, C. Technical and Economical Assessment of CO2 Capture-Based Ammonia Aqueous. *Processes* **2022**, *10*, 859. [CrossRef]
5. Czakiert, T.; Krzywanski, J.; Zylka, A.; Nowak, W. Chemical Looping Combustion: A Brief Overview. *Energies* **2022**, *15*, 1563. [CrossRef]
6. Hatta, N.S.M.; Aroua, M.K.; Hussin, F.; Gew, L.T. A Systematic Review of Amino Acid-Based Adsorbents for CO_2 Capture. *Energies* **2022**, *15*, 3753. [CrossRef]
7. Mendiara, T.; García-Labiano, F.; Abad, A.; Gayán, P.; de Diego, L.F.; Izquierdo, M.T.; Adánez, J. Negative CO_2 emissions through the use of biofuels in chemical looping technology: A review. *Appl. Energy* **2018**, *232*, 657–684. [CrossRef]
8. Tramošljika, B.; Blecich, P.; Bonefačić, I.; Glažar, V. Advanced Ultra-Supercritical Coal-Fired Power Plant with Post-Combustion Carbon Capture: Analysis of Electricity Penalty and CO_2 Emission Reduction. *Sustainability* **2021**, *13*, 801. [CrossRef]
9. Madejski, P.; Chmiel, K.; Subramanian, N.; Kuś, T. Methods and Techniques for CO_2 Capture: Review of Potential Solutions and Applications in Modern Energy Technologies. *Energies* **2022**, *15*, 887. [CrossRef]
10. Ziółkowski, P.; Badur, J.; Pawlak-Kruczek, H.; Niedzwiecki, L.; Kowal, M.; Krochmalny, K. A novel concept of negative CO_2 emission power plant for utilization of sewage sludge. In Proceedings of the 6th International Conference on Contemporary Problems of Thermal Engineering CPOTE 2020, Gliwice, Poland, 21–24 September 2020; pp. 531–542.
11. Ziółkowski, P.; Madejski, P.; Amiri, M.; Kuś, T.; Stasiak, K.; Subramanian, N.; Pawlak-Kruczek, H.; Badur, J.; Niedźwiecki, Ł.; Mikielewicz, D. Thermodynamic Analysis of Negative CO2 Emission Power Plant Using Aspen Plus, Aspen Hysys, and Ebsilon Software. *Energies* **2021**, *14*, 6304. [CrossRef]
12. Ziółkowski, P.; Głuch, S.; Ziółkowski, P.J.; Badur, J. Compact High Efficiency and Zero-Emission Gas-Fired Power Plant with Oxy-Combustion and Carbon Capture. *Energies* **2022**, *15*, 2590. [CrossRef]
13. Jacobs, H.R. Direct-Contact Condensation. In *Direct-Contact Heat Transfer*; Kreith, F., Boehm, R.F., Eds.; Springer: Berlin/Heidelberg, Germany, 1988; pp. 223–236. [CrossRef]
14. Desideri, U.; Di Maria, F. Simulation code for design and off design performance prediction of geothermal power plants. *Energy Convers. Manag.* **2000**, *41*, 61–76. [CrossRef]
15. Li, Y.; Klausner, J.F.; Mei, R. Performance characteristics of the diffusion driven desalination process. *Desalination* **2006**, *196*, 188–209. [CrossRef]
16. Ebrahimi, M.; Keshavarz, A.; Jamali, A. Energy and exergy analyses of a micro-steam CCHP cycle for a residential building. *Energy Build.* **2012**, *45*, 202–210. [CrossRef]
17. Marugán-Cruz, C.; Sánchez-Delgado, S.; Gómez-Hernández, J.; Santana, D. Towards zero water consumption in solar tower power plants. *Appl. Therm. Eng.* **2020**, *178*, 115505. [CrossRef]
18. Yan, M.; Ma, C.; Shen, Q.; Song, Z.; Chang, J. A novel lignite pre-drying system integrated with flue gas waste heat recovery at lignite-fired power plants. *Appl. Therm. Eng.* **2019**, *150*, 200–209. [CrossRef]
19. Zhao, X.; Fu, L.; Wang, X.; Sun, T.; Wang, J.; Zhang, S. Flue gas recovery system for natural gas combined heat and power plant with distributed peak-shaving heat pumps. *Appl. Therm. Eng.* **2017**, *111*, 599–607. [CrossRef]
20. Amann, J.M.; Kanniche, M.; Bouallou, C. Natural gas combined cycle power plant modified into an O_2/CO_2 cycle for CO_2 capture. *Energy Convers. Manag.* **2009**, *50*, 510–521. [CrossRef]
21. Garlapalli, R.K.; Spencer, M.W.; Alam, K.; Trembly, J.P. Integration of heat recovery unit in coal fired power plants to reduce energy cost of carbon dioxide capture. *Appl. Energy* **2018**, *229*, 900–909. [CrossRef]

22. Wei, M.; Zhao, X.; Fu, L.; Zhang, S. Performance study and application of new coal-fired boiler flue gas heat recovery system. *Appl. Energy* **2017**, *188*, 121–129. [CrossRef]

23. Prananto, L.A.; Juangsa, F.B.; Iqbal, R.M.; Aziz, M.; Soelaiman, T.A.F. Dry steam cycle application for excess steam utilization: Kamojang geothermal power plant case study. *Renew. Energy* **2018**, *117*, 157–165. [CrossRef]

24. Mati, A.; Buffi, M.; Dell'Orco, S.; Lombardi, G.; Ramiro, P.M.R.; Kersten, S.R.A.; Chiaramonti, D. Fractional Condensation of Fast Pyrolysis Bio-Oil to Improve Biocrude Quality towards Alternative Fuels Production. *Appl. Sci.* **2022**, *12*, 4822. [CrossRef]

25. Thongtip, T.; Aphornratana, S. An experimental analysis of the impact of primary nozzle geometries on the ejector performance used in R141b ejector refrigerator. *Appl. Therm. Eng.* **2017**, *110*, 89–101. [CrossRef]

26. Bencharif, M.; Nesreddine, H.; Perez, S.C.; Poncet, S.; Zid, S. The benefit of droplet injection on the performance of an ejector refrigeration cycle working with R245fa. *Int. J. Refrig.* **2020**, *113*, 276–287. [CrossRef]

27. Klausner, J.F.; Li, Y.; Mei, R. Evaporative heat and mass transfer for the diffusion driven desalination process. *Heat Mass Transf.* **2006**, *42*, 528–536. [CrossRef]

28. Wang, Y.; Hu, Y.; Huang, Q.; Cui, Y. Transient heat transfer study of direct contact condensation of steam in spray cooling water. *Trans. Tianjin Univ.* **2018**, *24*, 131–143. [CrossRef]

29. Hu, Y.; Wang, Y.; Huang, Q.; Cui, Y. Transient analysis of the steam-water direct contact condensation in the packed column. *Can. J. Chem. Eng.* **2018**, *96*, 404–413. [CrossRef]

30. Poškas, R.; Sirvydas, A.; Kulkovas, V.; Jouhara, H.; Poškas, P.; Miliauskas, G.; Puida, E. An Experimental Investigation of Water Vapor Condensation from Biofuel Flue Gas in a Model of Condenser, (2) Local Heat Transfer in a Calorimetric Tube with Water Injection. *Processes* **2021**, *9*, 1310. [CrossRef]

31. Negative CO2 Emission Gas Power Plant. Available online: https://nco2pp.mech.pg.gda.pl/en (accessed on 25 September 2022).

32. Madejski, P.; Kuś, T.; Subramanian, N.; Karch, M.; Michalak, P. The possibilities of carrying out numerical and experimental tests of jet type flow condensers for application in energy technologies. In Proceedings of the Wdzydzeanum Workshop on Fluid-Solid Interaction, Gdańsk, Poland, 5–10 September2021.

33. Madejski, P.; Karch, M.; Michalak, P.; Banasiak, K.; Kuś, T.; Subramanian, N. Measurement techniques and apparatus for experimental studies of flow processes in a system with a jet flow type condenser. In Proceedings of the XIV Multiphase Workshop and Summer School, Koszałkowo–Wieżyca k, Gdańsk, Poland, 2–4 September 2021.

34. Madejski, P.; Kuś, T.; Subramanian, N. Development of the analysis results for the possibility of using spray-ejector condensers for efficient operation in a negative CO_2 emission power plant cycle. In *Report WP1 Task 2*; POLNOR CCS nCO2PP: Kraków, Poland, March 2021.

35. Ziółkowski, P.; Stasiak, K.; Amiri, M.; Dąbrowski, P.; Mikielewicz, D. Completion of the PFD (Process Flow Diagram) along with detailed mass, momentum and energy balance of the negative CO_2 emission gas power plant (nCO2PP) based on steam gasification of sewage sludge, coupled with CCS. In *Report WP1 Task 1*; POLNOR CCS nCO2PP: Gdańsk, Poland, January 2021.

36. National Institute of Standards and Technology, Carbon Dioxide Tables. Available online: https://webbook.nist.gov/cgi/inchi/InChI%3D1S/CO2/c2-1-3 (accessed on 9 September 2022).

37. Helm, I.; Jalukse, L.; Leito, I. Measurement Uncertainty Estimation in Amperometric Sensors: A Tutorial Review. *Sensors* **2010**, *10*, 4430–4455. [CrossRef]

38. Chen, J.; Chen, C. Uncertainty Analysis in Humidity Measurements by the Psychrometer Method. *Sensors* **2017**, *17*, 368. [CrossRef]

39. Chen, L.-H.; Chen, J.; Chen, C. Effect of Environmental Measurement Uncertainty on Prediction of Evapotranspiration. *Atmosphere* **2018**, *9*, 400. [CrossRef]

40. Jing, H.; Quan, Z.; Zhao, Y.; Wang, L.; Ren, R.; Liu, Z. Thermal Performance and Energy Saving Analysis of Indoor Air–Water Heat Exchanger Based on Micro Heat Pipe Array for Data Center. *Energies* **2020**, *13*, 393. [CrossRef]

41. Ali, A.; Cocchi, L.; Picchi, A.; Facchini, B. Experimental Determination of the Heat Transfer Coefficient of Real Cooled Geometry Using Linear Regression Method. *Energies* **2021**, *14*, 180. [CrossRef]

42. Advisory Note No. 1. Uncertainties in Enthalpy for the IAPWS Formulation 1995 for the Thermodynamic Properties of Ordinary Water Substance for General and Scientific Use (IAPWS-95) and the IAPWS Industrial Formulation 1997 for the Thermodynamic Properties of Water and Steam (IAPWS-IF97). The International Association for the Properties of Water and Steam. Vejle, Denmark. August 2003. Available online: http://www.iapws.org/relguide/Advise1.pdf (accessed on 12 September 2022).

43. Wang, J.; Jia, C.-S.; Li, C.-J.; Peng, X.-L.; Zhang, L.-H.; Liu, J.-Y. Thermodynamic Properties for Carbon Dioxide. *ACS Omega* **2019**, *4*, 19193–19198. [CrossRef] [PubMed]

44. Michalski, L.; Eckersdorf, K.; Kucharski, J.; McGhee, J. *Temperature Measurement*, 2nd ed.; John Wiley & Sons Ltd.: Hoboken, NJ, USA, 2001.

45. Lee, T.-W. *Thermal and Flow Measurements*; CRC Press, Taylor & Francis Group, LLC: Boca Raton, FL, USA, 2008.

46. *IEC 60751:2022*; Industrial Platinum Resistance Thermometers and Platinum Temperature Sensors. International Electrotechnical Commission: Geneva, Switzerland, 2022.

47. Nicholas, J.V.; White, D.R. *Traceable Temperatures*; John Wiley & Sons, Ltd.: Chichester, UK, 2001.

48. *IEC 60584-1:2013*; Thermocouples-Part 1: EMF Specifications and Tolerances. International Electrotechnical Commission: Geneva, Switzerland, 2013.

49. Temperature Sensors—Product Catalogue. Limatherm Sensor. Available online: https://www.limathermsensor.pl/files/downloads/Catalogue_temperature_sensors_old.pdf (accessed on 25 September 2022).

50. Sandberg, R.J. *Temperature*; CRC Press LLC: Boca Raton, FL, USA, 2000.

51. Smith, C.L. *Basic Process Measurements*; John Wiley & Sons, Inc.: Hoboken, NJ, USA, 2009.
52. Zosel, J.; Oelßner, W.; Decker, M.; Gerlach, G.; Guth, U. The measurement of dissolved and gaseous carbon dioxide concentration. *Meas. Sci. Technol.* **2011**, *22*, 072001. [CrossRef]
53. Mendes, L.B.; Ogink, N.W.M.; Edouard, N.; Van Dooren, H.J.C.; Tinôco, I.D.F.F.; Mosquera, J. NDIR Gas Sensor for Spatial Monitoring of Carbon Dioxide Concentrations in Naturally Ventilated Livestock Buildings. *Sensors* **2015**, *15*, 11239–11257. [CrossRef]
54. Gerlach, G.; Guth, U.; Oelßner, W. *Carbon Dioxide Sensing: Fundamentals, Principles, and Applications*; Wiley-VCH Verlag GmbH & Co.: Weinheim, Germany, 2019.
55. Pedemonte, A.A.; Traverso, A.; Massardo, A.F. Experimental analysis of pressurised humidification tower for humid air gas turbine cycles. Part A: Experimental campaign. *Appl. Therm. Eng.* **2008**, *28*, 1711–1725. [CrossRef]
56. Pedemonte, A.A.; Traverso, A.; Massardo, A.F. Experimental analysis of pressurised humidification tower for humid air gas turbine cycles. Part B: Correlation of experimental data. *Appl. Therm. Eng.* **2008**, *28*, 1623–1629. [CrossRef]
57. Apanasevich, P.; Coste, P.; Ničeno, B.; Heib, C.; Lucas, D. Comparison of CFD simulations on two-phase Pressurized Thermal Shock scenarios. *Nucl. Eng. Des.* **2014**, *266*, 112–128. [CrossRef]
58. Mahood, H.B.; O Sharif, A.; Alaibi, S.; Hawkins, D.; Thorpe, R.B. Analytical solution and experimental measurements for temperature distribution prediction of three-phase direct-contact condenser. *Energy* **2014**, *67*, 538–547. [CrossRef]
59. Chen, X.; Tian, M.; Zhang, G.; Leng, X.; Qiu, Y.; Zhang, J. Visualization study on direct contact condensation characteristics of sonic steam jet in subcooled water flow in a restricted channel. *Int. J. Heat Mass Transf.* **2019**, *145*, 118761. [CrossRef]
60. Yang, X.P.; Liu, J.P.; Zong, X.; Chong, D.T.; Yan, J.J. Experimental study on the direct contact condensation of the steam jet in subcooled water flow in a rectangular channel: Flow patterns and flow field. *Int. J. Heat Fluid Flow* **2015**, *56*, 172–181. [CrossRef]
61. Xu, Q.; Guo, L.; Zou, S.; Chen, J.; Zhang, X. Experimental study on direct contact condensation of stable steam jet in water flow in a vertical pipe. *Int. J. Heat Mass Transf.* **2013**, *66*, 808–817. [CrossRef]
62. Hu, R.; Jin, Y.; Huang, Q.; Wang, Y.; Cui, Y. The effect of packing on direct contact evaporation in spray column. *Can. J. Chem. Eng.* **2017**, *95*, 2209–2220. [CrossRef]
63. Genić, S. Direct-contact condensation heat transfer on downcommerless trays for steam-water system. *Int. J. Heat Mass Transf.* **2006**, *49*, 1225–1230. [CrossRef]
64. Reddick, C.; Sorin, M.; Sapoundjiev, H.; Aidoun, Z. Effect of a mixture of carbon dioxide and steam on ejector performance: An experimental parametric investigation. *Exp. Therm. Fluid Sci.* **2018**, *92*, 353–365. [CrossRef]
65. Chen, X.; Tian, Z.; Guo, F.; Yu, X.; Huang, Q. Experimental investigation on direct-contact condensation of subatmospheric pressure steam in cocurrent flow packed tower. *Energy Sci. Eng.* **2022**, *10*, 2954–2969. [CrossRef]
66. Datta, P.; Chakravarty, A.; Saha, R.; Chaudhuri, S.; Ghosh, K.; Mukhopadhyay, A.; Sen, S.; Dutta, A.; Goyal, P. Experimental investigation on the effect of initial pressure conditions during steam-water direct contact condensation in a horizontal pipe geometry. *Int. Commun. Heat Mass Transf.* **2021**, *121*, 105082. [CrossRef]
67. Kwidzinski, R. Condensation heat and mass transfer in steam–water injectors. *Int. J. Heat Mass Transf.* **2021**, *164*, 120582. [CrossRef]
68. Zare, S.; Jamalkhoo, M.H.; Passandideh-Fard, M. Experimental Study of Direct Contact Condensation of Steam Jet in Water Flow in a Vertical Pipe With Square Cross Section. *Int. J. Multiph. Flow* **2018**, *104*, 74–88. [CrossRef]
69. Zong, X.; Liu, J.-P.; Yang, X.-P.; Yan, J.-J. Experimental study on the direct contact condensation of steam jet in subcooled water flow in a rectangular mix chamber. *Int. J. Heat Mass Transf.* **2015**, *80*, 448–457. [CrossRef]

Article

Multi-Model-Based Predictive Control for Divisional Regulation in the Direct Air-Cooling Condenser

Zhiling Luo [1,*] **and Qi Yao** [2]

[1] School of Control and Computer Engineering, North China Electric Power University, Beijing 102206, China
[2] Energy and Electricity Research Center, Jinan University, Zhuhai 519070, China; qiyao@jnu.edu.cn
* Correspondence: aloha.l@163.com

Abstract: Flow distortions caused by ambient wind can have complex negative effects on the performance of direct air-cooling condensers, which use air as their cooling medium. A control-oriented model of the direct air-cooling condenser model, considering fan volumetric effectiveness and plume recirculation rate, was developed, and its linearization model was derived. The influences of fan volumetric effectiveness and plume recirculation rate on backpressure were analyzed, and the optimal backpressure was calculated. To improve both the transient performance and steady-state energy saving of the condenser, a multi-model-based predictive control strategy was proposed to divisionally adjust the fan array. Four division schemes of the direct air-cooling fan array constituted the local models, and in each division scheme, axial fans were divided into three groups according to the wind direction: windward fans, leeward fans, and other fans. The simulation results showed that the turbine backpressure can be increased by 15 kPa under the influence of plume recirculation and the reduction of the fan volumetric efficiency. The fan division adjustment strategy can achieve satisfactory control performance with switching rules.

Keywords: backpressure; direct air-cooling condenser; divisional regulation; multi-model; plume recirculation

---

Citation: Luo, Z.; Yao, Q. Multi-Model-Based Predictive Control for Divisional Regulation in the Direct Air-Cooling Condenser. *Energies* **2022**, *15*, 4803. https://doi.org/10.3390/en15134803

Academic Editor: Pawel Madejski

Received: 17 May 2022
Accepted: 27 June 2022
Published: 30 June 2022

Publisher's Note: MDPI stays neutral with regard to jurisdictional claims in published maps and institutional affiliations.

1. Introduction

Direct air-cooling condensers are widely used in coal-fired and concentrated solar power plants because of their water-saving advantages [1]. As the specific heat capacity of air is smaller than that of water, the heat exchange efficiency is reduced, resulting in a significantly higher pressure in the direct air-cooling condenser than in the water-cooled condenser, which leads to a reduction in the turbine power output [2,3]. To improve system efficiency, it is important to understand the direct air-cooling condenser performance and study the relationship between condenser pressure and related variables.

Installed on an outdoor platform with a height of tens of meters, the performance of the direct air-cooling condenser is affected by factors such as turbine steam load ratio, fan speed, ambient temperature, wind speed, wind direction, and windscreens [4–7]. Flow distortion or disturbance caused by wind can negatively impact the performance of the condenser, resulting in complex fluctuations in fan inlet flow and temperature. Volumetric effectiveness and plume recirculation rate are used to evaluate the effect of the fan inlet flow and temperature fluctuations on direct air-cooling condenser performance, respectively. The crosswinds significantly reduce the volumetric effectiveness of windward fans, while winds along with the longitudinal lead to the increase in plume recirculation [8].

Fan volumetric effectiveness is defined as the ratio of the airflow through the fan with inlet disturbance to the airflow of the fan operating independently and without inlet disturbance, which is also called cluster factor in recent studies. Through scaled-down model experiments, Salta [9] found that the fan volumetric effectiveness of peripheral fans was always lower than that of internal fans, and presented the relationship between fan volumetric effectiveness and platform height, fan diameter, and walkway width. Fourie [10] derived

---

the correlation between fan volumetric effectiveness and crosswind speed. Through CFD simulation, Yang [11] presented that the aerodynamic characteristics of axial fans located at different positions of the fan array had different performances: the volumetric effectiveness of windward fans was lower than that of internal fans under the crosswind, while the volumetric effectiveness of leeward fans was higher, and the volumetric effectiveness value was affected by wind speed and direction.

Plume recirculation occurs when a part of the hot plume exiting the direct air-cooling condenser returns into the fan inlet [8]. The plume recirculation rate fluctuates with the wind direction and is relatively high when the wind direction is from the boiler direction, because the air temperature from the direction of the boiler is warmer, which reduces the heat exchange efficiency of the condenser. In operation, there have been cases of emergency shutdowns of power plants due to a crosswind in the direction of the boiler. Louw simulated the effects of various wind speeds and directions on a large air-cooling condenser with 386 air-cooling units and concluded that windward fans had the worst performance, and the edge fans were affected by plume recirculation [12].

In the research of modeling and optimization of the direct air-cooling condenser, numerical simulation and experimental research are the main methods. There are few studies about the development of control-oriented models, which have significant effects on improving power plant efficiency. Guo [13] established a dynamic model of the air-cooling condenser, analyzed the influence of fan face velocity and ambient temperature on backpressure, and applied it to the primary frequency regulation of a grid. Yang [14] combined the backpressure regulation of the direct air-cooling condenser with the boiler-turbine coordinated control system and achieved backpressure optimization and an accelerated turbine load response by controlling the fan speed. Zhang [15] developed a direct air-cooling condenser model based on the moving-boundary method and adopted feedforward compensation to suppress the backpressure disturbance caused by the ambient temperature variation. This model considered fan volumetric effectiveness but did not show its relationship to backpressure. None of these dynamic models reflected the negative effects of plume circulation.

In control-oriented modeling, the direct air-cooling system is usually simplified as one air-cooling module or one finned tube, and dozens of fans are regarded as a whole, which is adjusted uniformly. Studies on the performance of direct air-cooling condensers have suggested the necessity of fan divisional control to improve system efficiency. Liu [16] calculated the influence of operating conditions, fan speeds, and environmental parameters on the backpressure and coal consumption of direct air-cooling power plants through grey correlation analysis, which can be a reference for fan array divisional regulation. Li [17] simulated the flow field and temperature field of the direct air-cooling condenser, calculated the correlation between the rotational speed of each fan and the condenser pressure, and suggested adjusting the fans with a relatively higher correlation with condenser pressure is beneficial to reducing condenser pressure and improving efficiency. Huang [18,19] calculated optimal fan speed adjustment strategies under different meteorological conditions and proposed adjusting the fan speed in different blocks under different wind directions. Through scale-down experiments, we found that fan array divisional regulation can help save about 13% of fan energy consumption [20]. However, it is difficult to estimate model parameters and implement a control system, because few power plants are equipped with measurement points for volumetric effectiveness and plume recirculation rate systems. To overcome the problem of no measurement points, we designed a multi-model predictive control system that switched based on local model errors.

The contribution of this paper is to develop a dynamic direct air-cooling condenser model and propose a multi-model-based predictive control strategy for backpressure regulation. The novelties of this paper include: (1) establishing a control-oriented direct air-cooling condenser model, which considers volumetric effectiveness and the plume recirculation rate, and derives its linearization model; (2) the effects of volumetric effectiveness and plume recirculation rate on the dynamic response time and static gain of backpressure under various operating conditions are demonstrated; (3) the local model of the direct air-cooling condenser adopts fan divisional regulation schemes that change with the wind direction; (4) a model switching signal is designed to realize the switching of various fan divisional regulation schemes. The rest of this paper is organized as follows: Section 2 introduces the structure and the layout of the direct air-cooling condenser. Section 3 establishes and linearizes a dynamic model of an air-cooling unit of the condenser, validates the model, and analyzes the performance of the model. In Section 4, a multi-model predictive control strategy is proposed, and simulations are demonstrated. Finally, conclusions are drawn in Section 5.

2. Description of the System

The investigated 330 MW coal-fired power plant direct air-cooling condenser, located in Hebei Province, China, consists of 30 cooling units in five rows and six columns, arranged in an array shape. As shown in Figure 1, each cooling unit consists of an axial flow fan and hundreds of finned tube bundles. The finned tube bundles form an A-shaped structure. The exhaust steam from the turbine is sent to the finned tube bundles of the outdoor direct air-cooling condenser through the pipe. The cold air in the environment driven by the axial fan flows through the outer surface of the finned tubes and exchanges heat with the steam in order to condense the turbine exhaust steam into water. Part of the hot air leaving the air-cooling unit returns to the fan inlet under the influence of flow distortion and the ambient wind, which is called plume recirculation. The condensed water returns to the heat recovery system through the condensate pump. The pressure of the condenser is a key parameter that affects the power output of the coal-fired power plant. It needs to be optimized and regulated to a certain range. The designed parameters of the condenser are presented in Table 1.

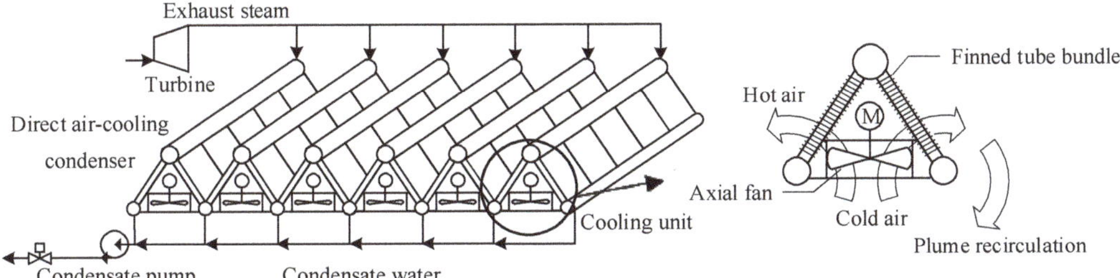

Figure 1. The schematic of the cold-end system.

Table 1. The Designed Parameters of the Direct Air-Cooling Condenser.

Parameter (Unit)	Value
Power plant nominal power (MW)	330
Exhaust steam mass flow rate (t/h)	741
Exhaust steam enthalpy (kJ/kg)	2531
Backpressure (kPa)	15
Ambient temperature (°C)	15
Fan rotational speed (r/min)	100
Fan nominal power (kW)	110
Fan front face area (m^2)	211
Fan face velocity (m/s)	2.46
Finned tube number of each cooling unit (-)	370
Finned tube length (m)	10.4
Finned tube mass (kg)	15.2
Finned tube combined specific heat capacity (kJ/(kg·K))	2.72
Finned tube wall density (kg/m^3)	7850
Finned tube cross-section area (m^2)	0.0033
Finned tube inner surface area (m^2)	2.06
Finned tube outer surface area (m2)	61.9
Finned tube inner surface heat transfer coefficient (W/(m^2·K))	7200
Finned tube outer surface heat transfer coefficient (W/(m^2·K))	35

Under the complex influence of ambient wind, the heat exchange efficiency of the air-cooling units located in different positions of the direct air-cooling condenser is significantly different. Figure 2 is the planer layout of the investigated coal-fired power plant, including the distribution of the direct air-cooling condenser, turbine house, and boiler house. Every circle represents a cooling unit. The local prevailing wind direction is 330°. The figure also shows the windward and leeward fans when the wind direction is 270°. The winds in the directions of 0° and 90° are crosswind, which significantly reduces the volumetric effectiveness of windward fans. The winds in the directions of 180° and 270° are longitudinal wind, which mainly leads to the increase in plume recirculation. The boiler is set in the downwind direction to avoid the effect of the warm crosswind in the direction of the boiler on the condenser.

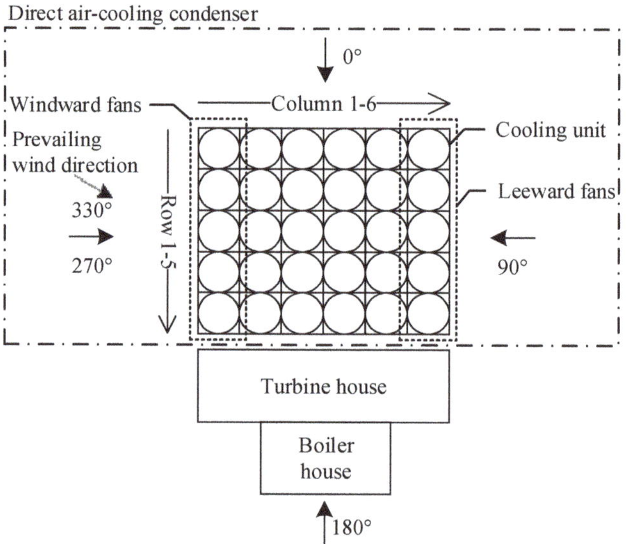

Figure 2. The planar layout of a direct air-cooling generating unit.

3. Model Development

This section presents the process model of an air-cooling unit, describing the dynamic effects of volumetric effectiveness, plume recirculation, and fan speed on the condenser, and the model is linearized and validated for control. Based on the air-cooling unit model established by Zhang [15], this paper originally describes the plume recirculation of the air-cooling unit and the derivation of the model linearization.

3.1. Process Model Development

With the conservation of energy and mass, the air-cooling unit can be described by a lumped parameter model with the moving-boundary approach [15].

The assumptions utilized in the derivation of the model include [21]: (1) the air-cooling unit is simplified into a long and thin horizontal tube; (2) there is only a two-phase region in the air-cooling unit; (3) the steam flows through the tube one-dimensionally; (4) the axial heat conduction and pressure drop are negligible. The schematic of the air-cooling unit model is shown in Figure 3.

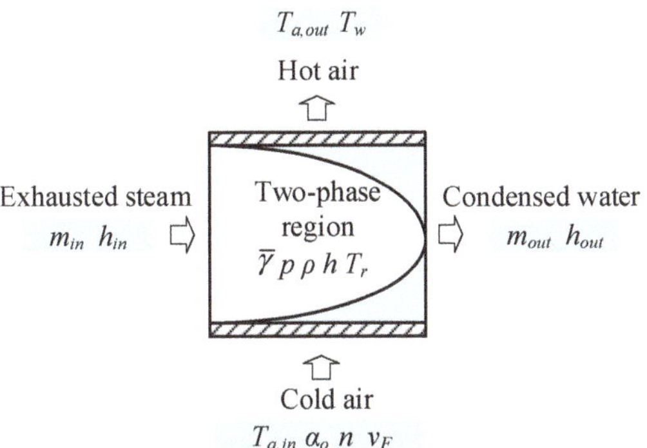

Figure 3. A schematic of the simplified air-cooling unit model.

1. Conservation of steam mass and energy

Mass conservation:

$$A_s L \frac{d\rho}{dt} = \dot{m}_{in} - \dot{m}_{out} \tag{1}$$

where ρ is the average density in the two-phase region; $\dot{m}$ is the steam mass flow rate; the subscript *in* and *out* represent the inlet and outlet of the condenser, respectively; A_s and L are the cross-section area and length of the finned tube, respectively.

Energy conservation:

$$A_s L \left(\frac{d\rho h}{dt} - \frac{dp}{dt} \right) = \dot{m}_{in} h_{in} - \dot{m}_{out} h_{out} + \alpha_i A_i (T_w - T_r) \tag{2}$$

where h and p are the enthalpy and condenser pressure in the two-phase region, respectively; α_i and A_i are the heat transfer coefficient and area of the inner surface of the finned tubes, respectively; T is temperature; the subscript w and r represent the finned tube wall and steam, respectively.

The density and enthalpy of the two-phase region are calculated as follows:

$$\rho = \overline{\gamma}\rho_g + (1 - \overline{\gamma})\rho_l \tag{3}$$

$$h = \frac{\overline{\gamma}\rho_g h_g + (1 - \overline{\gamma})\rho_l h_l}{\overline{\gamma}\rho_g + (1 - \overline{\gamma})\rho_l} \tag{4}$$

The subscript l and g are liquid and gas, respectively; $\overline{\gamma}$ is the two-phase mean void fraction, defined as the ratio of gas volume to two-phase region volume, which can be determined as [21]:

$$\overline{\gamma} = \frac{c}{(x_{out} - x_{in})(c - 1)^2} \ln \left[\frac{x_{in}(c - 1) - c}{x_{out}(c - 1) - c} \right] - \frac{1}{c - 1} \tag{5}$$

where $c = (\rho_g/\rho_l)^{2/3}$, x is the steam dryness, and the subscript *in* and *out* represent the inlet and outlet of the condenser, respectively.

By differentiating the variables, Equations (1), (2) and (5) can be expressed as:

$$A_s L \left[\frac{d\rho_l}{dp}(1 - \overline{\gamma}) + \frac{d\rho_g}{dp}\overline{\gamma} \right] \frac{dp}{dt} + A_s L (\rho_g - \rho_l) \frac{d\overline{\gamma}}{dt} = \dot{m}_{in} - \dot{m}_{out} \tag{6}$$

$$A_s L \left[\frac{d\rho_l h_l}{dp}(1 - \overline{\gamma}) + \frac{d\rho_g h_g}{dp}\overline{\gamma} - 1 \right] \frac{dp}{dt} + A_s L (\rho_g h_g - \rho_l h_l) \frac{d\overline{\gamma}}{dt} = \dot{m}_{in} h_{in} - \dot{m}_{out} h_{out} + \alpha_i A_i (T_w - T_r) \tag{7}$$

$$\frac{d\bar{\gamma}}{dt} = \frac{\partial\bar{\gamma}}{\partial p}\frac{dp}{dt} + \frac{\partial\bar{\gamma}}{\partial x_{in}}\left(\frac{\partial x_{in}}{\partial p}\frac{dp}{dt} + \frac{\partial x_{in}}{\partial h_{in}}\frac{dh_{in}}{dt}\right) + \frac{\partial\bar{\gamma}}{\partial x_{out}}\left(\frac{\partial x_{out}}{\partial p}\frac{dp}{dt} + \frac{\partial x_{out}}{\partial h_{out}}\frac{dh_{out}}{dt}\right) \tag{8}$$

2. Conservation of finned tube wall energy

$$C_{p,w}m_w\frac{dT_w}{dt} = \alpha_i A_i(T_r - T_w) + \alpha_o A_o(T_{a,in} - T_w) \tag{9}$$

where $C_{p,w}$ is the finned tube's combined specific heat capacity, which is calculated from the specific heat capacity of the fins and the tube wall [15]; m_w is the finned tube's mass; α_o and A_o are the heat transfer coefficient and area of the outer surface of the finned tube, respectively; T_a is air temperature; the subscripts *in* represents the inlet of the axial fan.

3. Conservation of air energy

$$C_{p,a}\rho_a v_F A_F(T_{a,in} - T_{a,out}) = \alpha_o A_o(T_{a,in} - T_w) \tag{10}$$

where $C_{p,a}$ and ρ_a are specific heat capacity and air density, respectively; v_F and A_F are face velocity and the front face area of the fan, respectively; $T_{a,out}$ is the air temperature of the fan outlet.

4. Air plume recirculation

The plume recirculation rate, which evaluates the proportion of air leaving the fan that returns to the fan inlet, is defined as [8]:

$$R = \frac{T_{a,in} - T_{a,e}}{T_{a,out} - T_{a,e}} \tag{11}$$

where $T_{a,e}$ is the air temperature of the environment. Combining (10) and (11), eliminating $T_{a,out}$, the relationship between $T_{a,in}$ and $T_{a,e}$ can be obtained:

$$T_{a,in} = \frac{(1 - R)C_{p,a}\rho_a v_F A_F T_{a,e} + R\alpha_o A_o T_w}{(1 - R)C_{p,a}\rho_a v_F A_F + R\alpha_o A_o} \tag{12}$$

5. Heat transfer coefficient

The heat transfer coefficient inside the finned tube is calculated with laminar condensation correlation [2]:

$$\alpha_i = 1.13\left(\frac{g\sin\varphi\rho^2\lambda^3\phi}{\mu(T_r - T_w)L}\right)^{\frac{1}{4}} \tag{13}$$

where Reynolds number $Re < 1600$, g is gravity acceleration, φ is the angle of the A-frame configuration, λ is the thermal conductivity, ϕ is wall heat flux density, and μ is the dynamic viscosity of the fluid.

The heat transfer coefficient outside the finned tube can be calculated with the following forced convection heat transfer correlation [2]:

$$\alpha_o = 0.19\frac{\lambda}{d_H}Re^{0.6} \tag{14}$$

where d_H is the hydraulic diameter, $Re = u_{max}d_H/v$ is the Reynolds number, $2 \times 10^3 < Re < 1.5 \times 10^4$, u_{max} is the airflow rate at the narrowest cross-section, and v is the kinematic viscosity of the air, which depends mainly on the temperature.

In a variable condition, an empirical formula can be used to calculate the heat transfer coefficient outside the finned tube [2]:

$$\alpha_o = \alpha_{o,0}\left(\frac{v_F v_0}{v_{F,0} v}\right)^{0.6} \tag{15}$$

where the subscript 0 represents the designed condition.

6. Axial fan

According to the fan similarity laws, axial fans have the following relationships between volumetric flow rate and power consumption:

$$\frac{V_F}{V_{F,0}} = \frac{n}{n_0} \tag{16}$$

$$\frac{E_F}{E_{F,0}} = \left(\frac{n}{n_0}\right)^3 \tag{17}$$

where V_F is the fan volumetric flow rate, E_F is the power consumption of an axial fan, n is the fan speed, and subscript 0 represents the nominal value.

Fan volumetric effectiveness, defined as the ratio of the actual airflow through the fan to that of the same fan operating without inlet disturbance, can be expressed as [9]:

$$\eta_F = \frac{V_{F,ac}}{V_F} \tag{18}$$

where $V_{F,ac}$ is the actual air flow rate of the fan. The fan face velocity can be calculated by:

$$v_F = \frac{V_{F,ac}}{A_F} \tag{19}$$

7. Exhaust steam mass flow and enthalpy

The exhaust steam mass flow rate can be calculated by:

$$\dot{m}_{in} = D_0 - \sum D_i \tag{20}$$

where D_0 is the main steam mass flow rate, and D_i is the steam mass flow rate from the ith extraction stage. Statistics show that the exhaust steam mass flow rate is proportional to the turbine steam load ratio β. Therefore, the exhaust steam mass flow rate can be obtained by polynomial fitting with the data provided by the manufacturer, and the expression is $\dot{m}_{in} = 634.1\beta + 52.44$.

Because turbine exhaust steam is wet and dryness is difficult to measure, and the expansion process of steam in the turbine is considered to be an isentropic expansion process, the interstage efficiency model is used to calculate the exhaust enthalpy.

$$h_{in} = h_j - \eta_T\left(h_j - \tilde{h}_{in}\right) \tag{21}$$

where h_j is the last extraction steam enthalpy, η_T is the turbine isentropic efficiency, and $\tilde{h}_{in}$ is the ideal enthalpy of the exhaust steam. The turbine isentropic efficiency is a function of the pressure ratio between the exhaust steam and last extraction steam, which can be expressed as a quadratic polynomial. The theoretical specific enthalpy of the exhaust steam can be obtained with the known exhaust steam pressure and last extraction steam theoretical entropy.

8. Governing equations

Combining the governing Equations (6)–(9) into a matrix form results in:

$$\dot{x} = D^{-1} f(x, u) \tag{22}$$

with state variables $x = \begin{bmatrix} p & \bar{\gamma} & h_{out} & T_w \end{bmatrix}^T$ and the manipulated variables $u = \begin{bmatrix} m_{in} & h_{in} & T_{a,e} & n \end{bmatrix}^T$, where

$$D = \begin{bmatrix} d_{11} & d_{12} & 0 & 0 \\ d_{21} & d_{22} & 0 & 0 \\ 0 & 0 & 0 & d_{34} \\ d_{41} & d_{42} & d_{43} & 0 \end{bmatrix}, \quad f = \begin{bmatrix} \dot{m}_{in} - \dot{m}_{out} \\ \dot{m}_{in}h_{in} - \dot{m}_{out}h_{out} + \alpha_i A_i (T_w - T_r) \\ \alpha_i A_i (T_r - T_w) + \alpha_o A_o (T_{a,in} - T_w) \\ -\dfrac{\partial \bar{\gamma}}{\partial x_{in}} \dfrac{\partial x_{in}}{\partial h_{in}} \dfrac{dh_{in}}{dt} \end{bmatrix}.$$

The elements in matrix D are given in Appendix A.

The linearized model of the direct air-cooling condenser can be used to investigate the dynamic behavior at an operating point (x_0, u_0). Hence, the model of the direct air-cooling condenser can be linearized as follows:

$$\Delta \dot{x} = A \Delta x + B \Delta u \tag{23}$$

where $A = \left[D|_{(x_0,u_0)} \right]^{-1} \left. \dfrac{\partial f(x,u)}{\partial x} \right|_{(x_0,u_0)}$, $B = \left[D|_{(x_0,u_0)} \right]^{-1} \left. \dfrac{\partial f(x,u)}{\partial u} \right|_{(x_0,u_0)}$, and the elements in matrix A and B are given in Appendix A.

3.2. Optimum Condenser Pressure

Turbine backpressure is a parameter that affects the operating economy of the power plant, because when the backpressure increases, the generating power output decreases. The regulation of the axial fan speed is the main way to adjust the backpressure. However, increasing the fan speed to reduce the backpressure is not an ideal operation strategy, since the energy consumed by the axial fans cannot be ignored. When the difference between the turbine power output and the axial fan power is the largest, the fan speed and the corresponding backpressure are optimal. Therefore, this optimization problem is formulated with the following fitness function:

$$max \; \Delta E = E_T - E_F \tag{24}$$

where ΔE is the net power, and E_T is the turbine power output.

According to the relationship between the relative deviation of the turbine power output and the backpressure provided by the manufacturer, the turbine power output under operating conditions can be calculated as [14]:

$$E_T = E_{T,0} \beta (1 + \zeta) \tag{25}$$

where $E_{T,0}$ is the nominal turbine power output, β is the turbine steam load ratio, ζ is the turbine power relative deviation, $\zeta = f(\beta)(p - p_0)$, and $f(\beta) = 0.0034\beta - 0.0064$ is the slope of the backpressure–turbine power increment curve, obtained by polynomial fitting with data provided by the manufacturer.

3.3. Model Validation

The dynamic and linearized models were validated with a fan speed step test of the investigated direct air-cooling condenser of the 330 MW power plant. Before the step test, the fan inlet temperature and face velocity of an edge fan and an internal fan were measured, and the average plume recirculation and fan volumetric effectiveness were calculated. In the step test, the rotation speeds of all fans were adjusted simultaneously. First, the fan rotation speeds were changed from 66 r/min to 44 r/min at $t = 15$ s, and then from 44 r/min to 65 r/min at $t = 500$ s. Figure 4 shows the experimental and simulation results of backpressure variation caused by step changes in fan speed. Comparing the four curves, the established model showed a similar dynamic response and steady-state to the actual system. The root mean square error (RMSE) of the nonlinear model was 0.11 kPa, and the RMSE of the linearized model was 0.22 kPa, when R = 0.02 and $\eta = 0.85$. When R = 0 and $\eta = 1$, the RMSE of the nonlinear model was 1.17 kPa, which indicated that the calculated backpressure is lower than the actual value when the plume recirculation and the fan volumetric effectiveness are not considered.

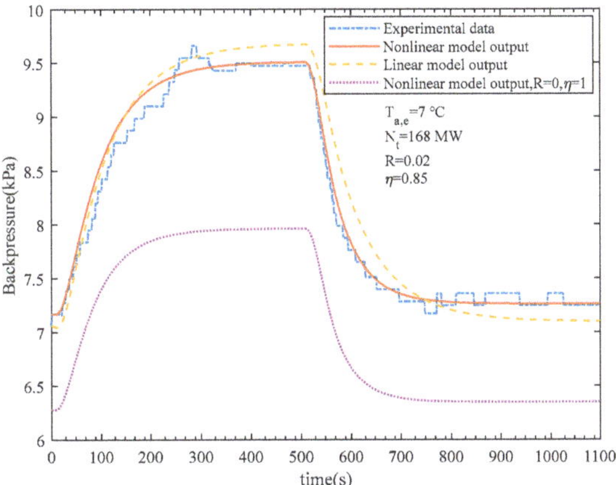

Figure 4. Experimental and simulation results of fan speed step test.

3.4. Model Performance Analysis

1. Step test response

Figure 5 shows the backpressure response at different fan volumetric effectiveness and plume recirculation rates when the fan speed was reduced by 10 r/min under nominal operating conditions. When the operating conditions were unchanged, the condenser performance was significantly affected by the ambient wind. The fan volumetric effectiveness and plume recirculation rate affected both the dynamic response time and static gain of backpressure, and the plume recirculation rate has a particularly significant effect on the static gain of backpressure.

2. Steady-state performance

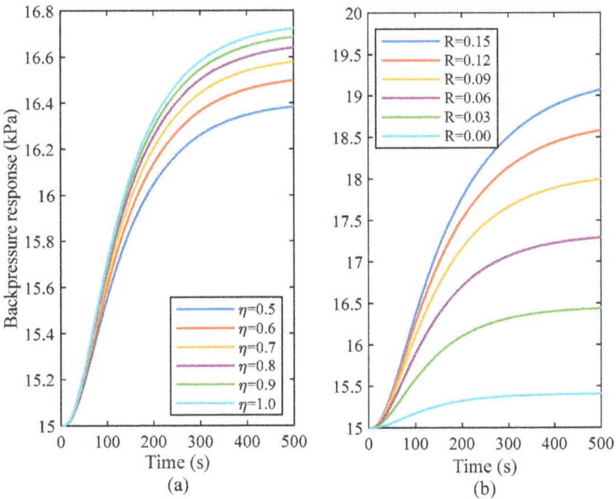

Figure 5. Response of backpressure to fan speed step test: (**a**) backpressure at different fan volumetric effectiveness; (**b**) backpressure at different plume recirculation rates.

Figures 6–8 show the steady-state performance of the direct air-cooling condenser model under different operating conditions. Figures 6 and 7 show that the backpressure increased with increasing ambient temperature and an increasing turbine steam load ratio, and with decreasing fan rotational speed, which is consistent with the model performance in the literature. The influence of ambient temperature on backpressure is significant, and adjusting the fan speed is an effective way to regulate backpressure.

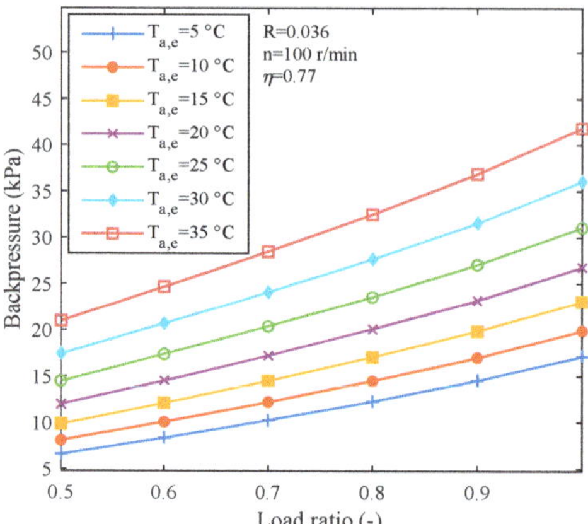

Figure 6. Backpressure at different ambient temperatures and load ratios.

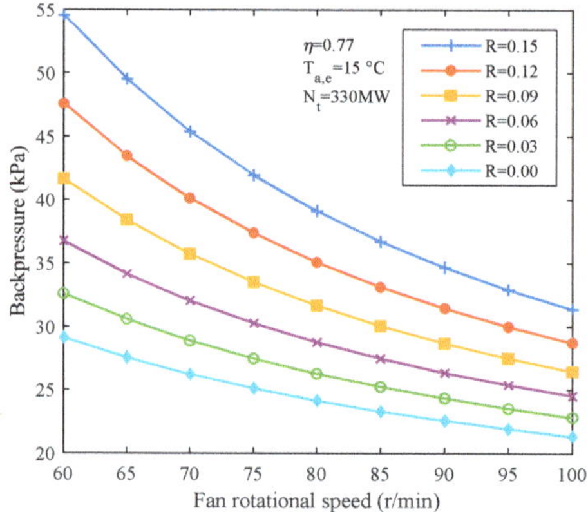

Figure 7. Backpressure at different fan speeds and plume recirculation rates.

In Figure 7, the condenser pressure increases with the plume recirculation rate. As the fan speed decreases, the influence of the change of the plume recirculation rate on the backpressure increases. The effect of plume recirculation on backpressure is similar to that of ambient temperature on backpressure, because plume recirculation causes an increase in fan inlet temperature. Compared with the design operating point of the condenser ($R = 1$, and $n = 100$ r/min), when the plume recirculation rate of the fan rose to 0.15, the backpressure increased by 10 kPa.

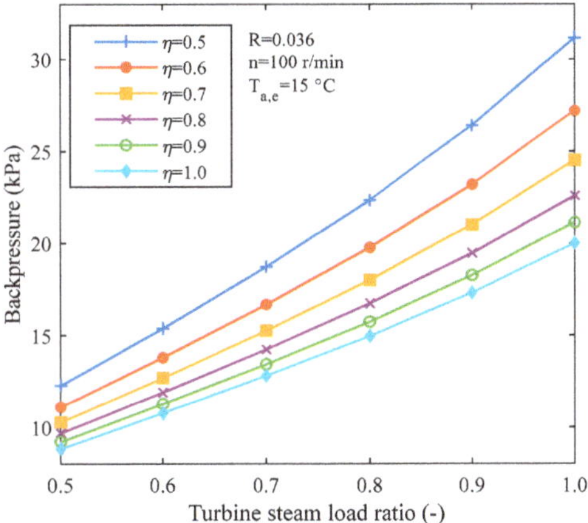

Figure 8. Backpressure at different load ratios and fan volumetric effectiveness.

In Figure 8, the condenser pressure increases as the fan volumetric efficiency decreases. When the turbine steam load ratio increases, the effect of fan volumetric effectiveness on backpressure increases. The fan face velocity is decreased under the disturbance of ambient wind, which is similar to the effect of fan speed reduction on backpressure. Compared with the design operating point of the condenser ($\eta = 1$, and $N_T = 330$ MW), when the volumetric effectiveness decreased by 0.5, the backpressure rose by 11 kPa.

3.   Optimum backpressure

The optimum backpressure under different fan volumetric effectiveness and plume recirculation rates were calculated according to Equation (24) and are shown in Figure 9, when $T_{a,e} = 15\ ^\circ$C and $N_T = 231$ MW. The optimal backpressure decreases as the volumetric effectiveness increases, and decreases as the plume recirculation rate decreases. Compared with the design operation point of the condenser ($\eta = 1$, and $R = 0$), when $\eta = 0.5$ and $R = 0.15$, the optimum backpressure of the condenser increased by nearly 15 kPa.

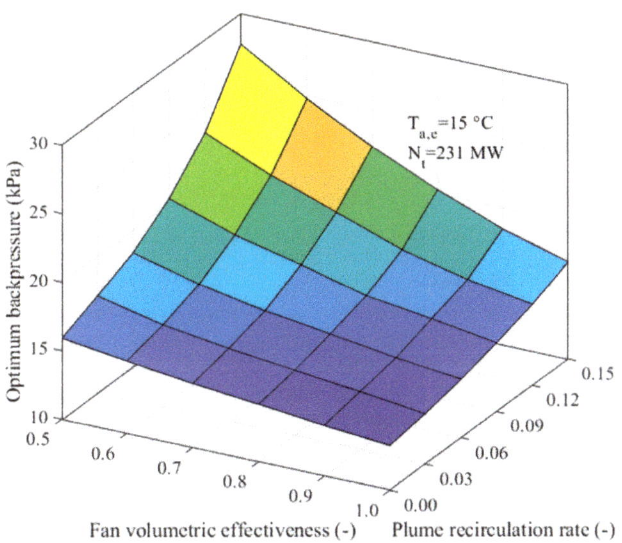

Figure 9. Optimum backpressure at different plume recirculation rates and fan volumetric effectiveness.

It was found that under different operating conditions, the fan volumetric effectiveness and plume recirculation rate have a significant impact on backpressure, which further affects the economy of the power plant. The performance explains why a sudden burst of ambient wind, especially from the direction of the boiler, can cause a rapid increase in the backpressure of the direct air-cooling power plant, resulting in an emergency shutdown.

4. Control Methodology

To overcome the negative effects of flow distortion or disturbances caused by ambient wind on the performance of the direct air-cooling condenser, a multi-model predictive control strategy was proposed to improve the control performance and economy.

4.1. Global Structure

The structure of the multi-model predictive controller for backpressure regulation is depicted in Figure 10. The difference between the output of each local model $y_m(k)$ and the output of the plant $y(k)$ is used to generate a model switching signal $\delta(k)$. The switching signal is sent to the model optimization and model prediction modules, respectively. The difference between the optimal set-point value generated by the optimization $r(k)$ and the predicted output of the selected model $y_p(k)$ is transmitted to the controller to calculate the control variable $u(k)$. Backpressure optimization is introduced in Section 3.2. The following sections present the local models, model prediction, control objective function, and switching rules.

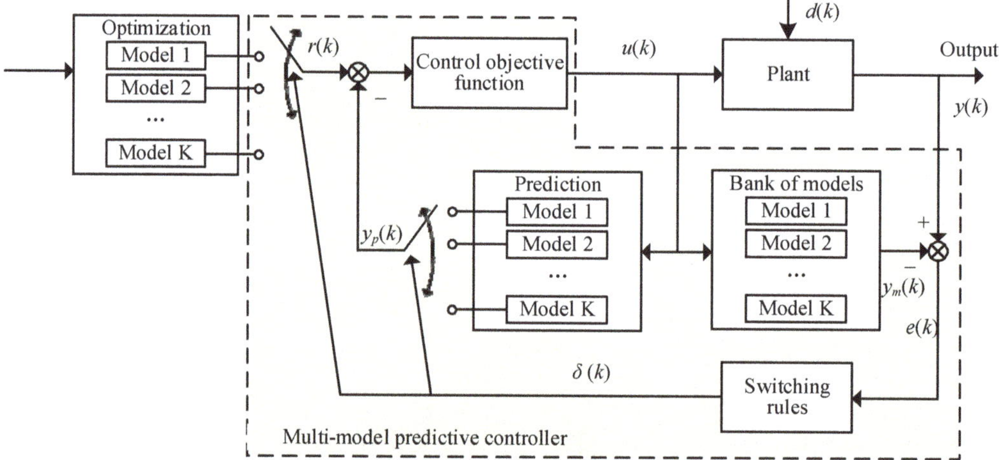

Figure 10. The structure of the multi-model predictive controller for the direct air-cooling condenser.

4.2. Bank of Local Model

At present, dozens of air-cooling units in the condenser are controlled as one unit, but the fans in different positions of the condenser have different performances under the ambient wind. Referring to the divisional control suggestions presented by previous studies [7,16–18], four division schemes were proposed for the direct air-cooling array according to the wind direction, as shown in Figure 11. The axial fans were divided into three groups according to the wind direction: windward fans, leeward fans, and other fans. Each group of fans was represented by an air-cooling unit model established in Equation (23) and adjusted separately. The average backpressure of all cooling units was taken as the direct air-cooling condenser backpressure.

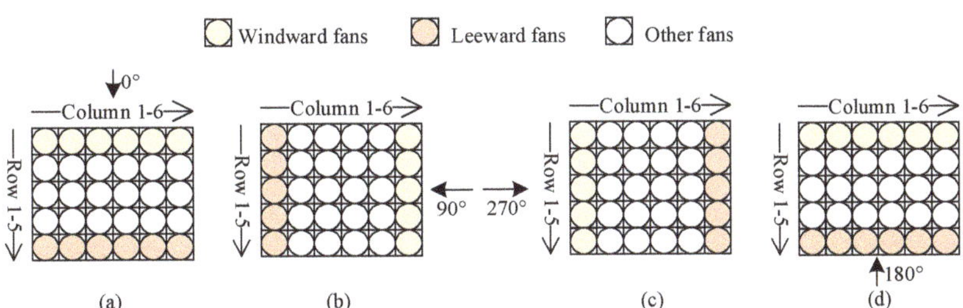

Figure 11. Schematic of divisional regulation of fan array under different wind directions: (**a**) in wind direction 0°; (**b**) in wind direction 90°; (**c**) in wind direction 270°; (**d**) in wind direction 180°.

When the wind direction is 90°, the influence of ambient wind on the airflow of the direct air-cooling condenser is axisymmetric with that of when the wind direction is 270°. Therefore, this paper shows the controller performance by switching between the three models in Figure 11a,c,d. When the ambient wind speed was 6 m/s, the plume recirculation rate and volumetric effectiveness were measured by field experiments and scaled-down model experiments, respectively. The establishment of the scaled-down fan array model and experiments are introduced in [20]. The parameters are shown in Table 2. The plume recirculation caused by ambient wind gradually decreased in the order of wind direction angles of 180°, 270°, and 0°. The plume recirculation rate and wind direction have a strong correlation at the same wind speed, and therefore, this characteristic is utilized in model switching.

Table 2. The Parameters of Local Models.

Wind Direction (°)	Average Plume Recirculation Rate (-)	Volumetric Effectiveness (-)		
		Windward Fans	Leeward Fans	Other Fans
0	0.036	0.55	0.85	0.78
270	0.06	0.5	0.97	0.79
180	0.15	0.55	0.85	0.78

4.3. Model Predictive Control

Model predictive control (MPC) is an effective, advanced control technique for constrained multiple-input multiple-output (MIMO) control problems in the process industry. The brief idea of MPC is as follows:

At every sampling instant k, a process model is used to predict the current and future values of the output variables with current measurements:

$$x(k+1) = Ax(k) + Bu(k)$$
$$y(k) = Cx(k) \tag{26}$$

The future output can be estimated as:

$$y(k+P) = C\left(A^P x(k) + \sum_{j=0}^{P-1} A^j B u_{P-1-j}\right) \tag{27}$$

Let $y_P(k) = [y(k+1)^T, \cdots, y(k+P)^T]^T$ be the predicted values for the output and $u_M(k) = [u(k+1)^T, \cdots, u(k+M-1)^T]^T$ be the future manipulated input, where the subscripts P and M are the prediction and control horizon, respectively. The manipulated input is held constant after the M control moves.

The constraints on the input and output are:

$$u_{\min} \leq u_M(k) \leq u_{\max}$$
$$y_{\min} \leq y_P(k) \leq y_{\max} \tag{28}$$

where the subscripts min and max stand for lower and upper bounds, respectively.

The future manipulated variables are calculated by optimizing an objective function so that a sequence of predicted output reaches the set-point in an optimal manner:

$$\min J(k) = \|y_P(k) - r(k)\|^2_{Q_y} + \|\Delta u_M(k)\|^2_{Q_u} \tag{29}$$

where Q_y and Q_u are the symmetric positive definite weight matrixes, and $r(k)$ is the set-point generated by the optimization. With the receding horizon approach, although a set of manipulated input is calculated, only the first move is implemented. Then, a new sequence of control is calculated at the next sampling time, with new measurements.

4.4. Model Switching Rules

When the environment or operating conditions of the system change, it is expected that the controller can quickly switch to the local model that most fits the current system state. A function of error is used as a criterion to evaluate which local model is optimal [22]:

$$\delta(k) = \arg \min_{i \in 1,2,\ldots,K} e_i^2(k) + \int_0^k e^{-\lambda(k-t)} e_i^2(t) dt \tag{30}$$

where $\delta(k)$ is the switch signal, $e_i(k) = y(k) - y_{m,i}(k)$ $(i = 1, \ldots, K)$ is the ith model output error at time k, and λ is a forgetting factor. The criteria of local models are calculated at every sampling time, the local model with the smallest criterion is adopted as the best fit model, and the switch signal is sent to the optimization and model prediction modules. An appropriate forgetting factor is beneficial to avoid frequent switching. A forgetting factor with a small value can effectively avoid unnecessary switching, while a forgetting factor with a large value is conducive to the rapid response of the model.

4.5. Simulation Results and Discussion

When the ambient wind direction changed, the simulation results of the direct air-cooling condenser using the multi-model controller are as shown in Figure 12. The tracking performance of the multi-model predictive controller and the traditional predictive controller are compared.

The operating condition in $t = 0$ s was: $N_T = 231$ MW, $T_{a,e} = 15$ °C, $p = 13.5$ kPa, $n = 100$ r/min, and R = 0.036. During the simulation: (1) At $t = 0$ s, the set-point of backpressure was increased to 14.4 kPa, which was the optimal backpressure. As the model fit the plant, there was no difference between the two controllers. (2) At $t = 80$ s, the plume recirculation rate increased from 0.036 to 0.15, resulting in an increment in condenser pressure, deviating from the set-point. With the proposed method, the prediction model was switched from model 1 to model 3, and the disturbance was quickly suppressed. (3) At $t = 150$ s, the set-point of backpressure was increased to 17.5 kPa, which was the optimal backpressure calculated according to model 3. The multi-model predictive controller acted rapidly to track the changed set-point, whereas the controller with a single prediction model experienced large overshoots and fluctuations, which were caused by the model mismatch. (4) At $t = 300$ s, the plume recirculation rate decreased to 0.1 There was no perfectly matching prediction model, but according to the switching rule, model 2 was the closest prediction model. The controller also provided a satisfactory effect. (5) At $t = 400$ s, the set-point of backpressure was decreased to 14.9 kPa, which was the optimal backpressure calculated according to model 2.

It can also be seen that while the divisional regulation of fans was adopted, the operating speeds of the three groups of fans were different, because the volumetric effectiveness of the fans was reflected in the models. When the backpressure set-point value was high, the rotation speed of the windward fan and the leeward fan was relatively large, which was the expected control result.

In general, the proposed multi-model predictive control strategy for the direct air-cooling condenser has satisfactory control performance, can quickly switch the prediction model, and provide the optimal backpressure reference value according to the prediction model.

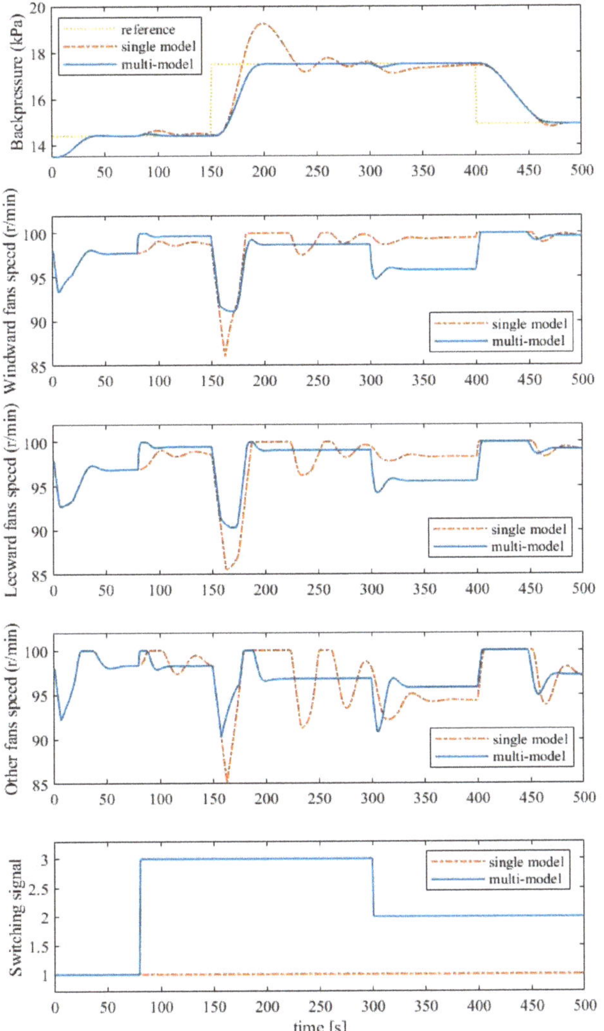

Figure 12. Tracking performance of the multi-model predictive controller.

5. Conclusions

A control-oriented direct air-cooling condenser model was established considering the fan volumetric effectiveness and the plume recirculation rate. The following conclusions were drawn:

(1) The model shows the influence of fan volumetric effectiveness and plume recirculation on the dynamic response and steady state of backpressure under various operating conditions.

(2) According to the wind direction, the fan array is divided into three groups: windward fans, leeward fans, and other fans, which are adjusted separately. At a certain wind speed, there is a strong correlation between wind direction and plume recirculation rate, and this relationship is utilized in model switching.

(3) A multi-model-based predictive control strategy for the direct air-cooling condenser is proposed. The analysis shows that applying the control strategy improves the robustness and performance of the system.

Author Contributions: Z.L.: conceptualization, methodology, investigation, simulation, and writing—original draft; Q.Y.: resources, data curation, formal analysis, and writing—review and editing. All authors have read and agreed to the published version of the manuscript.

Funding: This research was funded by the National Natural Science Foundation of China, grant number 51821004.

Conflicts of Interest: The authors declare no conflict of interest.

Appendix A

This section provides elements in matrix D of Equation (22) and matrix A and B of Equation (23). The thermodynamic properties and partial derivatives of steam and water can be calculated by a library based on the IAPWS IF-97.

$$D = \begin{bmatrix} d_{11} & d_{12} & 0 & 0 \\ d_{21} & d_{22} & 0 & 0 \\ 0 & 0 & 0 & d_{34} \\ d_{41} & d_{42} & d_{43} & 0 \end{bmatrix} \tag{A1}$$

$$d_{11} = A_s L \left(\frac{d\rho_l}{dp}(1 - \bar{\gamma}) + \frac{d\rho_g}{dp}\bar{\gamma} \right) \tag{A2}$$

$$d_{12} = A_s L \left(\rho_g - \rho_l \right) \tag{A3}$$

$$d_{21} = A_s L \left[\frac{d\rho_l h_l}{dp}(1 - \bar{\gamma}) + \frac{d\rho_g h_g}{dp}\bar{\gamma} - 1 \right] \tag{A4}$$

$$d_{22} = A_s L \left(\rho_g h_g - \rho_l h_l \right) \tag{A5}$$

$$d_{34} = C_w m_w \tag{A6}$$

$$d_{41} = \left(\frac{\partial \bar{\gamma}}{\partial p} + \frac{\partial \bar{\gamma}}{\partial x_{in}}\frac{\partial x_{in}}{\partial p} + \frac{\partial \bar{\gamma}}{\partial x_{out}}\frac{\partial x_{out}}{\partial p} \right) \tag{A7}$$

$$\frac{\partial \bar{\gamma}}{\partial p} = \frac{\partial \bar{\gamma}}{\partial c}\frac{dc}{dp} \tag{A8}$$

$$\frac{\partial \bar{\gamma}}{\partial c} = \frac{1}{x_{out} - x_{in}}\left(\frac{1}{(1-c)^2} - \frac{2c}{(c-1)^3} \right) ln\left[\frac{x_{in}(c-1) - c}{x_{out}(c-1) - c} \right]$$
$$+ \frac{c}{(x_{out} - x_{in})(c-1)^2}\left[\frac{x_{in} - 1}{x_{in}(c-1) - c} - \frac{x_{out} - 1}{x_{out}(c-1) - c} \right] + \frac{1}{(c-1)^2} \tag{A9}$$

$$\frac{dc}{dp} = \frac{2}{3}\left(\frac{\rho_g}{\rho_l} \right)^{-\frac{1}{3}}\left(\frac{1}{\rho_l}\frac{d\rho_g}{dp} - \frac{\rho_g}{\rho_l^2}\frac{d\rho_l}{dp} \right) \tag{A10}$$

$$\frac{\partial \bar{\gamma}}{\partial x_{in}} = \frac{c}{(x_{out} - x_{in})^2(c-1)^2}ln\left[\frac{x_{in}(c-1) - c}{x_{out}(c-1) - c} \right]$$
$$+ \frac{c}{(x_{out} - x_{in})(c-1)^2}\frac{c-1}{x_{in}(c-1) - c} \tag{A11}$$

$$\frac{\partial \overline{\gamma}}{\partial x_{out}} = \frac{-c}{(x_{out}-x_{in})^2(c-1)^2} ln\left[\frac{x_{in}(c-1)-c}{x_{out}(c-1)-c}\right] \\ -\frac{c}{(x_{tout}-x_{in})(c-1)^2}\frac{c-1}{x_{out}(c-1)-c}$$ (A12)

$$\frac{\partial x}{\partial p} = \frac{-1}{h_g - h_l}\left[(1-x)\frac{dh_l}{dp} + x\frac{dh_g}{dp}\right]$$ (A13)

$$d_{42} = -1$$ (A14)

$$d_{43} = \frac{\partial \overline{\gamma}}{\partial x_{out}}\frac{\partial x_{out}}{\partial h_{out}}$$ (A15)

$$\frac{\partial x}{\partial h} = \frac{1}{h_g - h_l}$$ (A16)

$$A = \left[D|_{(x_0,u_0)}\right]^{-1}\frac{\partial f(x,u)}{\partial x}\bigg|_{(x_0,u_0)}$$ (A17)

$$\frac{\partial f(x,u)}{\partial x} = \begin{bmatrix} 0 & 0 & 0 & 0 \\ f_{x,21} & 0 & f_{x,23} & f_{x,24} \\ f_{x,31} & 0 & 0 & f_{x,34} \\ 0 & 0 & 0 & 0 \end{bmatrix}$$ (A18)

$$f_{x,21} = -\alpha_i A_i \frac{dT_r}{dp}$$ (A19)

$$f_{x,23} = -\dot{m}_{out}$$ (A20)

$$f_{x,24} = \alpha_i A_i$$ (A21)

$$f_{x,31} = \alpha_i A_i \frac{dT_r}{dp}$$ (A22)

$$f_{x,34} = -\alpha_i A_i - \alpha_o A_o + \alpha_o A_o \frac{\partial T_{a,in}}{\partial T_w}$$ (A23)

$$\frac{\partial T_{a,in}}{\partial T_w} = \frac{R\alpha_o A_o}{(1-R)C_{p,a}\rho_a v_F A_F + R\alpha_o A_o}$$ (A24)

$$B = \left[D|_{(x_0,u_0)}\right]^{-1}\frac{\partial f(x,u)}{\partial u}\bigg|_{(x_0,u_0)}$$ (A25)

$$\frac{\partial f(x,u)}{\partial u} = \begin{bmatrix} f_{u,11} & 0 & 0 & 0 \\ f_{u,21} & f_{u,22} & 0 & 0 \\ 0 & 0 & f_{u,33} & f_{u,34} \\ 0 & 0 & 0 & 0 \end{bmatrix}$$ (A26)

$$f_{u,11} = 1$$ (A27)

$$f_{u,21} = h_{in}$$ (A28)

$$f_{u,22} = \dot{m}_{in}$$ (A29)

$$f_{u,33} = \alpha_o A_o \frac{\partial T_{a,in}}{\partial T_{a,e}}$$ (A30)

$$\frac{\partial T_{a,in}}{\partial T_{a,e}} = \frac{(1-R)C_{p,a}\rho_a v_F A_F}{(1-R)C_{p,a}\rho_a v_F A_F + R\alpha_o A_o}$$ (A31)

$$f_{u,34} = A_o\left[(T_{a,in} - T_w)\frac{\partial \alpha_o}{\partial v_F} + \alpha_o\left(\frac{\partial T_{a,in}}{\partial v_F} + \frac{\partial T_{a,in}}{\partial \alpha_o}\frac{\partial \alpha_o}{\partial v_F}\right)\right]\frac{\partial v_F}{\partial n}$$ (A32)

$$\frac{\partial T_{a,in}}{\partial \alpha_o} = \frac{R(1-R)A_o C_{p,a}\rho_a v_F A_F(T_w - T_{a,e})}{\left((1-R)C_{p,a}\rho_a v_F A_F + R\alpha_o A_o\right)^2}$$ (A33)

$$\frac{\partial T_{a,in}}{\partial v_F} = \frac{R(1-R)\alpha_o A_o C_{p,a}\rho_a A_F(T_{a,e}-T_w)}{\left((1-R)C_{p,a}\rho_a v_F A_F + R\alpha_o A_o\right)^2} \tag{A34}$$

$$\frac{\partial \alpha_o}{\partial v_F} = 0.6\alpha_{o,0}v_{F,0}{}^{-0.6}v_F{}^{-0.4} \tag{A35}$$

$$\frac{\partial v_F}{\partial n} = \frac{\eta_F v_{F,0}}{n_0} \tag{A36}$$

References

1. Huang, C.; Hou, H.; Hu, E.; Yu, G.; Peng, H.; Yang, Y.; Wang, L.; Zhao, J. Performance Maximization of a Solar Aided Power Generation (SAPG) Plant with a Direct Air-Cooled Condenser in Power-Boosting Mode. *Energy* **2019**, *175*, 891–899. [CrossRef]
2. Liu, J.; Hu, Y.; Zeng, D.; Wang, W. Optimization of an Air-Cooling System and Its Application to Grid Stability. *Appl. Therm. Eng.* **2013**, *61*, 206–212. [CrossRef]
3. Du, X.; Liu, L.; Xi, X.; Yang, L.; Yang, Y.; Liu, Z.; Zhang, X.; Yu, C.; Du, J. Back Pressure Prediction of the Direct Air Cooled Power Generating Unit Using the Artificial Neural Network Model. *Appl. Therm. Eng.* **2011**, *31*, 3009–3014. [CrossRef]
4. Marincowitz, F.S.; Owen, M.; Muiyser, J. Experimental Investigation of the Effect of Perimeter Windscreens on Air-Cooled Condenser Fan Performance. *Appl. Therm. Eng.* **2019**, *163*, 114395. [CrossRef]
5. Li, X.; Wang, N.; Wang, L.; Kantor, I.; Robineau, J.-L.; Yang, Y.; Maréchal, F. A Data-Driven Model for the Air-Cooling Condenser of Thermal Power Plants Based on Data Reconciliation and Support Vector Regression. *Appl. Therm. Eng.* **2018**, *129*, 1496–1507. [CrossRef]
6. Li, X.; Wang, N.; Wang, L.; Yang, Y.; Maréchal, F. Identification of Optimal Operating Strategy of Direct Air-Cooling Condenser for Rankine Cycle Based Power Plants. *Appl. Energy* **2018**, *209*, 153–166. [CrossRef]
7. He, W.F.; Chen, J.J.; Han, D.; Wen, T.; Luo, L.T.; Li, R.Y.; Zhong, W.C. Numerical Analysis from the Rotational Speed Regulation within the Fan Array on the Performance of an Air-Cooled Steam Condenser. *Appl. Therm. Eng.* **2019**, *153*, 352–360. [CrossRef]
8. Duvenhage, K.; Kröger, D.G. The Influence of Wind on the Performance of Forced Draught Air-Cooled Heat Exchangers. *J. Wind. Eng. Ind. Aerodyn.* **1996**, *62*, 259–277. [CrossRef]
9. Salta, C.A.; Kröger, D.G. Effect of Inlet Flow Distortions on Fan Performance in Forced Draught Air-Cooled Heat Exchangers. *Heat Recovery Syst. CHP* **1995**, *15*, 555–561. [CrossRef]
10. Fourie, N.; van der Spuy, S.J.; von Backström, T.W. Simulating the Effect of Wind on the Performance of Axial Flow Fans in Air-Cooled Steam Condenser Systems. *J. Therm. Sci. Eng. Appl.* **2015**, *7*, 21011. [CrossRef]
11. Yang, L.J.; Du, X.Z.; Zhang, H.; Yang, Y.P. Numerical Investigation on the Cluster Effect of an Array of Axial Flow Fans for Air-Cooled Condensers in a Power Plant. *Sci. Bull.* **2011**, *56*, 2272–2280. [CrossRef]
12. Louw, F.G. *Performance Trends of a Large Air-Cooled Steam Condenser during Windy Conditions*; Stellenbosch University: Stellenbosch, South Africa, 2011.
13. Guo, Y.; Zhang, D.; Wan, J.; Yu, D. Influence of Direct Air-Cooled Units on Primary Frequency Regulation in Power Systems. *IET Gener. Transm. Distrib.* **2017**, *11*, 4365–4372. [CrossRef]
14. Yang, T.; Wang, W.; Zeng, D.; Liu, J.; Cui, C. Closed-Loop Optimization Control on Fan Speed of Air-Cooled Steam Condenser Units for Energy Saving and Rapid Load Regulation. *Energy* **2017**, *135*, 394–404. [CrossRef]
15. Zhang, Y.; Zhang, F.; Shen, J. On the Dynamic Modeling and Control of the Cold-End System in a Direct Air-Cooling Generating Unit. *Appl. Therm. Eng.* **2019**, *151*, 373–384. [CrossRef]
16. Liu, L.; Du, X.; Xi, X.; Yang, L.; Yang, Y. Experimental Analysis of Parameter Influences on the Performances Ofdirect Air Cooled Power Generating Unit. *Energy* **2013**, *56*, 117–123. [CrossRef]
17. Li, J.; Bai, Y.; Li, B. Operation of Air Cooled Condensers for Optimised Back Pressure at Ambient Wind. *Appl. Therm. Eng.* **2018**, *128*, 1340–1350. [CrossRef]
18. Huang, W.; Chen, L.; Wang, W.; Yang, L.; Du, X. Cooling Performance Optimization of Direct Dry Cooling System Based on Partition Adjustment of Axial Flow Fans. *Energies* **2020**, *13*, 3179. [CrossRef]
19. Huang, W.; Chen, L.; Yang, L.; Du, X. Energy-Saving Strategies of Axial Flow Fans for Direct Dry Cooling System. *Energies* **2021**, *14*, 3176. [CrossRef]
20. Luo, Z.; Liu, J.; Huusom, J.K. Energy-Efficient Operation of a Direct Air-Cooled Condenser Based on Divisional Regulation. *Int. J. Refrig.* **2021**, *132*, 233–242. [CrossRef]
21. Qiao, H.; Laughman, C.R.; Aute, V.; Radermacher, R. An Advanced Switching Moving Boundary Heat Exchanger Model with Pressure Drop. *Int. J. Refrig.* **2016**, *65*, 154–171. [CrossRef]
22. Zhu, H.; Shen, J.; Lee, K.Y.; Sun, L. Multi-Model Based Predictive Sliding Mode Control for Bed Temperature Regulation in Circulating Fluidized Bed Boiler. *Control Eng. Pract.* **2020**, *101*, 104484. [CrossRef]

Article

Water Level Control in the Thermal Power Plant Steam Separator Based on New PID Tuning Method for Integrating Processes

Goran S. Kvascev * and Zeljko M. Djurovic

School of Electrical Engineering, University of Belgrade, 11120 Beograd, Serbia
* Correspondence: kvascev@etf.bg.ac.rs; Tel.: +381-11-3218-437

Abstract: The paper presents an analysis of water-level control in a thermal power plant (TPP) steam separator. This control structure is vital for the entire plant's stable, reliable, and efficient operation. This process belongs to processes with an integrator because it concerns a level-control issue, and the control variable is the feedwater flow. Said industrial processes are challenging to control and apply standard methods for tuning the PID controller, so a new procedure has been proposed. A procedure for tuning a PID controller for integrating processes is proposed based on the IFOPDT model, obtained from the wide step response of the process. Based on the process parameters estimated, the tuning of the controller is proposed. Results from the TPP TEKO-B2 (350 MW) are presented as an experimental verification. Compared with standard tuning methods, better results are achieved in the form of rise time and disturbance elimination rate. A significantly less risky and faster experiment for parameter estimation and controller tuning is also obtained. In addition, one adjustable parameter is provided to select the relation between performance and robustness. This method can be applied to various industrial processes with an integrator.

Keywords: thermal powerplant steam separator; tuning PD/PID; level control; integrating process; IFOPDT model

Citation: Kvascev, G.S.; Djurovic, Z.M. Water Level Control in the Thermal Power Plant Steam Separator Based on New PID Tuning Method for Integrating Processes. *Energies* **2022**, *15*, 6310. https://doi.org/10.3390/en15176310

Academic Editor: Pawel Madejski

Received: 24 July 2022
Accepted: 25 August 2022
Published: 29 August 2022

Publisher's Note: MDPI stays neutral with regard to jurisdictional claims in published maps and institutional affiliations.

1. Introduction

Thermal power plants are one of the most complex technological processes in terms of the production process complexity and the number of system components, including different technological processes, sensors, and actuators. Thermal power plants represent a significant source of electrical energy, and the reliability and efficiency of the entire system are essential for maintaining a stable supply chain to end consumers. This paper shall include the analysis of one of the critical thermal power plant subsystems, i.e., the water-level control subsystem in the steam separator of the Serbian TEKO B2 Drmno thermal power plant, with a nominal power of 350 MW. The implemented cascade control structure will be presented, and the process analysis will demonstrate that it is a process with an integrator. Control of such processes is significantly more difficult than processes without an integrator, so a novel method for tuning the PID controller parameters is proposed for efficient high-performance control. This paper is the result of many years of research in the theoretical domain but also is based on the practical implementation of control structures in more than 10 thermal power plants in Serbia with a power range from 100 MW to 650 MW.

During the design of the control, when the block reconstruction was carried out, the application of different control structures was considered. By analyzing the existing solutions, the present state of the industry [1], and the future system tuning and maintenance, a decision was made to use only PID controller-based structures. Considering recent analyses and comparisons of different types of controllers, taking into account the implementation options [2–4], we concluded that the right decision was made. Application of H_∞ and a model-based predictive controller (MPC) requires an excellent knowledge of the process model, which is quite challenging to provide in this branch of industry. On the other hand, in case there is some change or disturbance in the system that can cause instability in the controller operation, the PID controller is easy to stabilize by changing the proportional gain. In contrast, for modern controller types, it is not so simple—the controller must be redesigned, which requires significant time and the presence of control engineers to redesign the controller.

With the conclusion that PID structures shall be used for control, a decision was made that it was necessary to perform a good tuning of the PID parameters to achieve good performance of the closed-loop system. Certain classical controller-tuning methods could not be implemented due to the inadequate experiments they required since it was a process with an integrator. Thus, the use of the relay-feedback method for determining the controller parameters was proposed, which was a good proposal, considering the available research [5]. The application of said method was proposed by Rotac [6] back in 1961, and since then, many different modifications and improvements have also been proposed [7–11]. However, based on the pulse response-based method [12], it was suggested to carry out further research and application options for a simpler and shorter experiment at a high-risk plant.

At the beginning of the 20th century, the first PID controller forms appeared, initiated by the practical need for precise navigation of large ships. Ever since, one has been able to find dozens of different schemes in the literature for both tuning the parameters of said controllers and characterizing the controlled process. The book by [13] discloses over 400 rules for controller tuning for different process types, including the step response-based procedure [14], but for processes without integrators. All of this indicates that none of the said structures is universal and that each process class requires special treatment. This paper describes a procedure that is demanding from two points of view. First, processes with an integrator are discussed, implying that the critical experiment from which the process parameters are read cannot last too long. Simultaneously, there are significant limitations on the control signal. In addition, the procedure is suitable for industrial processes, representing a different problem from the constant presence of various types of disturbances and non-negligible measurement noise. The authors believe that our experience, which becomes available to a wider community of electrical and mechanical engineers through this work, represents the missing segment in the range of various techniques for PID controller tuning. At the same time, we are sure that the presented results will arouse great interest among engineers who are in charge of designing and controlling industrial facilities, such as the water supply system in thermal energy blocks.

The water-level controlling process in steam separators is an extremely demanding design task for the following reasons. It is an integrator process translating the level from the minimum to the maximum in a very short amount of time. Therefore, the duration of every critical experiment from which the descriptive parameters of the process should be derived must be concise. On the other hand, the increment of the nominal control participating in the critical experiment must be minimal and carefully calculated. Then the measurement noise can be comparable to the useful signal. Considering the high pressure prevailing in the steam separator, one that is over 200 bar; the highly fluctuating boundary between steam and water; and the indirect level measurement by the pressure at the bottom of the separator, the measurement signal is very dynamic with a significant presence of measurement noise and measurement uncertainty. Finally, the water level in the steam separator is one of the safety protection signals at the thermal power plant, due to which the superior protection system, in a very short time interval measured in seconds, switches off the thermo-energy block according to the emergency procedure. In other words, the design requirements are rigorous, do not allow significant overshoot or undershoot, and require a fast response and elimination of disturbances. Therefore, this process was an inspiring motive for designing a new procedure for setting the PID controller parameters for processes with integral action.

The paper will demonstrate that the authors have been successful and have proven the initial hypotheses, namely, that it is possible to perform a suitable identification of the process parameters and, based on them, set the PID controller for the water-level control in the separator.

The paper has two significant contributions: (1) a novel method for tuning the process with an integrator, wide pulse-response tuning, is presented, and (2) the application of the water-level control procedure in the steam separator of the thermal power plant has been demonstrated. The efficiency of this new method, the simplicity of tuning, and the improvement of the performance over classic methods, is proven. Section 2 describes the steam separator-level control process at the TEKO B2 thermal power plant in Serbia, and the proposed and implemented control structure. Section 3 illustrates the new procedure for the PID controller tuning and theoretical analysis, and Section 4 presents experimental results and the process verification described in Section 2. Section 5 includes a discussion of experimental results and guidelines for further development, and Section 6 is the conclusion.

2. Process Description

Thermal power plants are the largest generators of electricity in Serbia, contributing to more than 65% of the overall power supply. As such, their operational efficiency and stability need to be maximized. Particular emphasis is placed on reliable long-term operation in terms of negotiated delivery commitments, operation per design criteria for energy efficiency, and longevity of the facility.

The paper addresses the control of the water-level steam separator (drum) in thermal power plant boilers [15–17]. A boiler is a unit in which the chemical energy of fossil fuel is converted into heat energy as steam. Figure 1a shows the basic structure of a steam boiler. Mills, usually six or seven, crush and grind coal, and then the controlled mixture of coal and preheated air is fed to a furnace via a system of ducts. In parallel, the oxygen needed for combustion is provided by an air-supply fan. The air is preheated to enforce combustion on the way to the boiler. Temperatures inside the boiler are as high as 1400 °C.

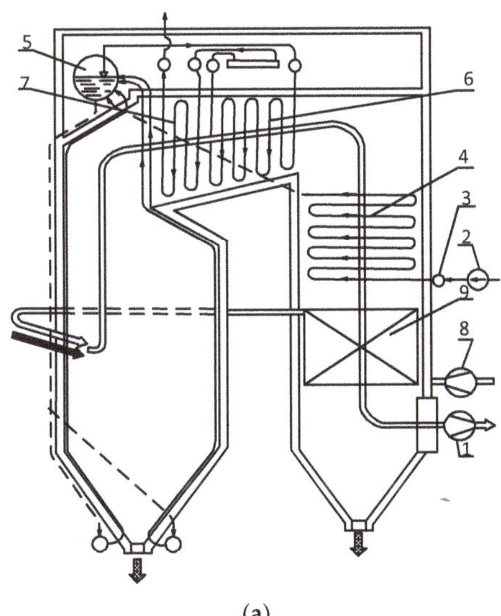

(a)

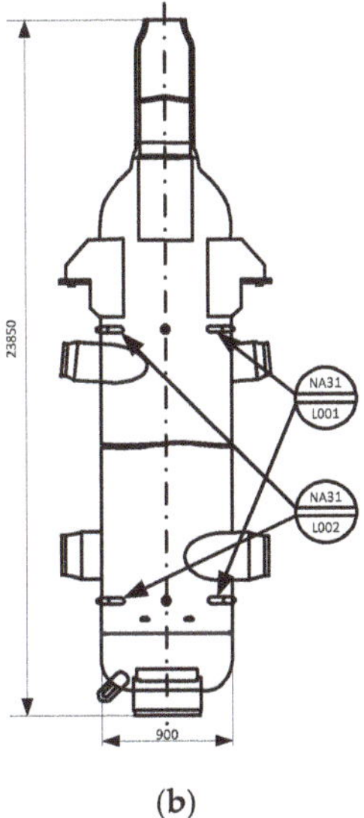

(b)

Figure 1. (**a**) Schematic view of a typical boiler: 1—exhaust fan, 2—feedwater pumps, 3—main feedwater valve, 4—economizer, 5—steam separator (drum), 6—primary preheater, 7—secondary preheater, 8—air supply fan, 9—air preheater; (**b**) schematic of a steam drum.

Feedwater pumps deliver preheated water to the steam drum via an economizer. Then additional pumps discharge the water into the system of pipes, where multi-stage heating takes place inside the boiler and the water is converted into steam. The steam separator also removes residual drops of water from the steam. The steam is then delivered to a multi-stage superheater, where it is heated to about 540 °C at a nominal pressure (usually 165–175 bars) before it leaves the boiler. The superheated steam flows to the turbine.

Specifically, at the TEKO B2 Unit of the Kostolac thermal power plant, nominal installed power is 350 MW, the diameter of the steam separator is 0.9 m, its height is about 24 m (Figure 1b), and it has a vertical orientation. Even a slight water-level variation inside the steam drum results in noticeable steam pressure fluctuations and affects the technical conditions of the process. If the water level is too high, emergency relief valves open to remove excess, but this measure reduces the unit's operational efficiency. However, if the water level is too low, after a short time, the boiler shutdown procedure is initiated automatically to protect the piping and installation from overheating. As a result, precise control of the water level is very important for stable and efficient work of the entire powerplant.

The feedwater subsystem is shown in Figure 2, which represents a realistic illustration of the SCADA system of the TEKO B2 Unit of the thermal power plant. Figure 2 shows the feedwater tank, three feedwater pumps (two of which are in operation, and one is spare), the main feedwater valve, and the steam drum (separator). Figure 2 shows all measured quantities in this subsystem: temperatures, levels, pressures, speed of pumps, actuator positions, and flows. Water from the feedwater tank is pumped via two pumps under high pressure (approx. 200 bar) into the steam separator (drum). Flow regulation is achieved through hydraulic couplings (manufactured by Voith) that receive the signal from the corresponding flow controllers, depending on the level in the separator. Then the feed water passes through the main feedwater valve, which keeps the pumps within the authorized operating mode, based on the pump protection Q-H curve, and reaches the steam separator.

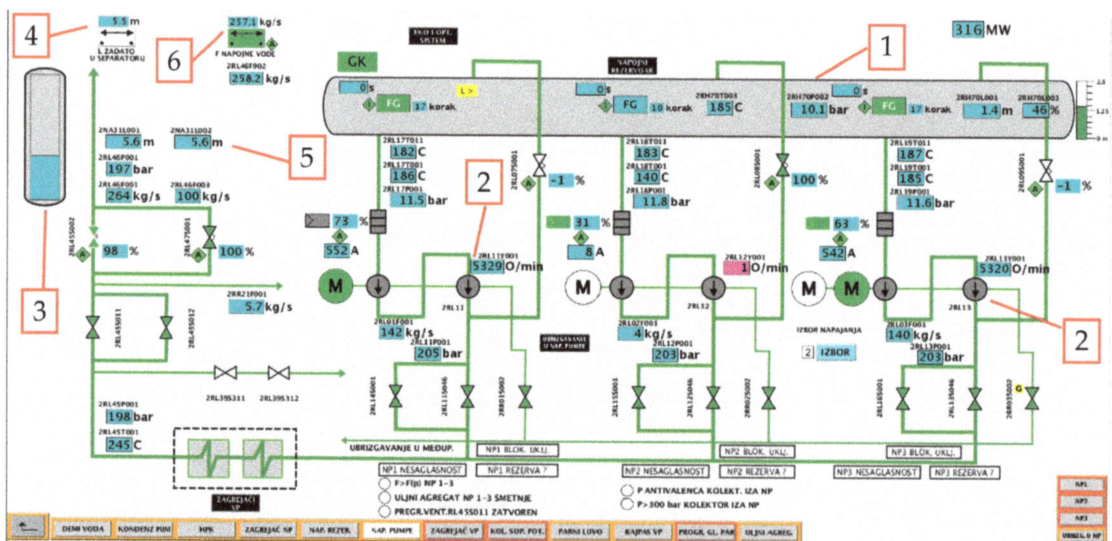

Figure 2. TE Kostolac B2 Unit, 350 MW nominal electric power—SCADA view of feedwater subsystem. 1—feedwater tank, 2—feedwater pumps, 3—steam separator (drum), 4—setpoint of water level, 5—measured water level in steam drum (separator), 6—calculated value (control variable) of feedwater flow from main PID controller.

The control structure is shown in Figure 3, which is a cascade-level control. With the mode selector different modes can be chosen, as well as PID control and two modes for parameter estimation for controller tuning. This will be explained later in Section 4. The main level controller is a PID controller (value of mode selector set to 1) that regulates the water level in the steam drum (separator), and its control signal is the set feedwater flow rate for each pump. This regulator is basically the PD controller, and the integral action has been added to eliminate disturbances and maintain feedwater flow control—nominal operating mode. The set flow rate is passed to the subordinate (slave) PID controllers for each pump. Said controllers have the task of regulating the set flow for each pump (RL31, RL32, RL33) in operation via adequate hydraulic couplings (Voith's). Their output signal is the Voith's set load percentage, in the range of 0–100%. This control structure is chosen to manage multiple actuators (pumps) and distribute the controller's functions into a level and flow control. The maximum performance of the control system is achieved as well as a high speed when changing the operating and spare pumps.

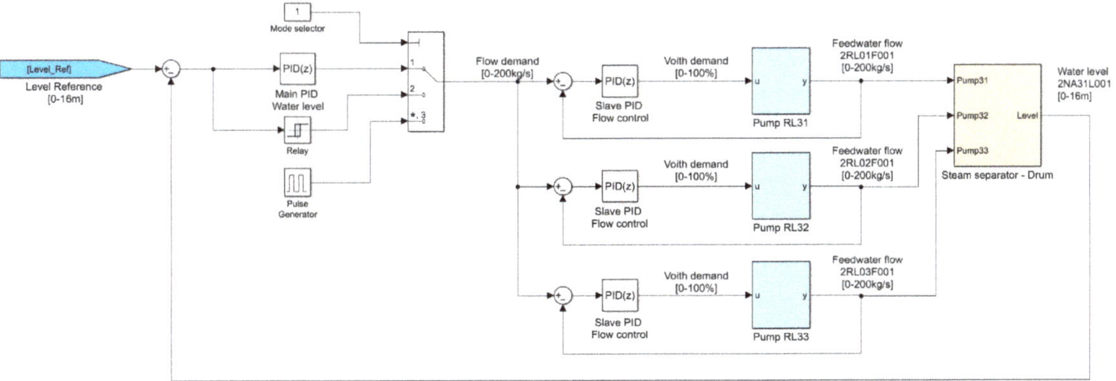

Figure 3. TE Kostolac B2 Unit, 350 MW nominal electric power—proposed PID cascade control design for level control of steam separator, with the operation mode selector: nominal mode—PID controller (mode 1), relay experiment (mode 2), wide pulse (mode 3).

The water level in the steam separator (drum) depends on the water flow to the drum and the steam flow from the drum. Since an integrating effect is inherent in the process, the main controller must be adapted for process control with an integrator. The application of classic methods for the PID controller tuning is either not feasible due to the impossibility of conducting appropriate experiments or inferior performance. The settings obtained in this way lead to control systems that do not meet the efficient operation criteria for said subsystem. Therefore, it is necessary to resort to tuning the PID controller for controlling processes with an integral effect. To achieve maximum performance, ease of tuning, and the length of the experiment for obtaining the control parameters, a new method based on wide pulse response tuning (WPRT) is proposed, as presented in the following section.

3. New PID Controller Tuning Procedure for Integrating Processes—Wide Pulse Response Tuning (WPRT)

To identify parameters of the IFOPDT model (integrating first order plus dead time) represented by $G(s) = \frac{K}{s(Ts+1)}e^{-\tau s}$, it is necessary to determine the following parameters:

K—gain;
T—dominant time constant;
τ—transport delay.

For step excitation, the response of process with the integrator constantly increases or decreases depending on the sign of process gain, and theoretically is not bounded. In practical applications, the system's output at some moment reaches a hardware limit due to plant constraints, limitations, or the protection system. Because of that, step response tuning does not apply to such plants and cannot be implemented in industrial practice.

If a process with only an integrator is observed, $G_1(s) = K/s$; as long as there is a constant input, the output of the process will rise. It will immediately stop rising at the end of step excitation, unlike the IFOPDT process output that does not stop increasing in the moment. This can be explained by the following: Steady-state time for the FOPDT processes is $(3 \div 5)T + \tau$ after every excitation change. Therefore, if the IFOPDT process $G(s)$ output is observed, part of it that represents FOPDT $(e^{-\tau s}/(Ts + 1))$ has an impact on the output $(3 \div 5)T + \tau$ after every step change of the input during a wide pulse experiment. This means that at the end of pulse excitation, $\frac{e^{-\tau s}}{Ts+1}$, which is part of $G(s)$, shall reach a new steady state after $(3 \div 5)T + \tau$, and that is one reason why the output will continue to rise after the end of wide pulse excitation.

Transfer function $G(s)$, as shown in Figure 4, can be divided into two processes, connected in series $G_{m1}(s) = K/s$ and $G_{m2}(s) = e^{-\tau s}/(Ts + 1)$.

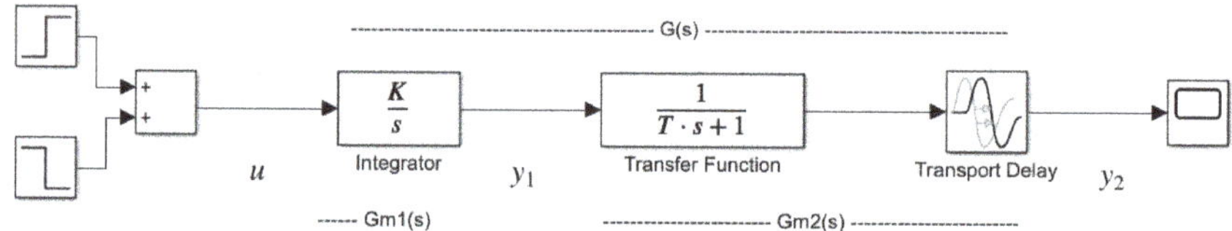

Figure 4. Ideal integrating first-order plus dead-time (IFOPDT) process $G(s) = \frac{K}{s(Ts+1)}e^{-\tau s}$.

If there is a constant step excitation u, y_1 (response of $G_{m1}(s)$) will be a function with a constant slope ($G_{m1}(s)$ is an integrator with gain). As for $G_{m2}(s)$, it will need $(3 \div 5)T + \tau$ to reach a new steady state, so y_2, which is the output of $G_{m2}(s)$ and, at the same time, the output of the entire process $G(s)$, will be a pure ramp function after $(3 \div 5)T + \tau$ time.

Now the second part of the IFOPDT model can be observed. y_1 is a ramp excitation for the process $G_{m2}(s)$. After reaching the ramp signal on y_2, which means that $G_{m2}(s)$ has reached steady state, pulse excitation is suspended, e.g., the process input has previous values before the experiment is started. Then, y_1 reaches the last value before excitation is stopped, and it maintains that value. As for $G_{m2}(s)$, whose input is y_1, the above mentioned process represents step excitation with a value of the difference of y_1 and y_2 (further Δy) at the end of the pulse. The y_2 signal will increase its value until $G_{m2}(s)$ reaches a new steady state (which is the value of y_1 at the time of the change).

The main idea is to determine the dynamic parameters of the IFOPDT model, τ and T, by measuring the time required for the process to reach 10% and 63% of the response. Model gain K can be obtained as the ratio of a total change of output y and the area of wide pulse excitation A. This will be discussed in detail in the next two subsections.

In industrial practice, we often encounter two process types, those with no or with negligibly small transport delay or those that have a significant transport delay. Therefore, to determine the parameters of the IFOPDT model, two separate cases must be considered. After the conducted open-loop experiment with wide step excitation, the process's output is observed. If there is a change in the output in a short time interval, we can talk about a process with no or negligibly small transport delay. If the output change does not occur during a short amount of time, then the case of a process with a significant transport delay must be considered.

3.1. Process with Small or No Transport Delay

If a process with small or no transport delay is considered (as shown in Figure 5), the parameters of the IFOPDT model can be obtained by the following procedure:

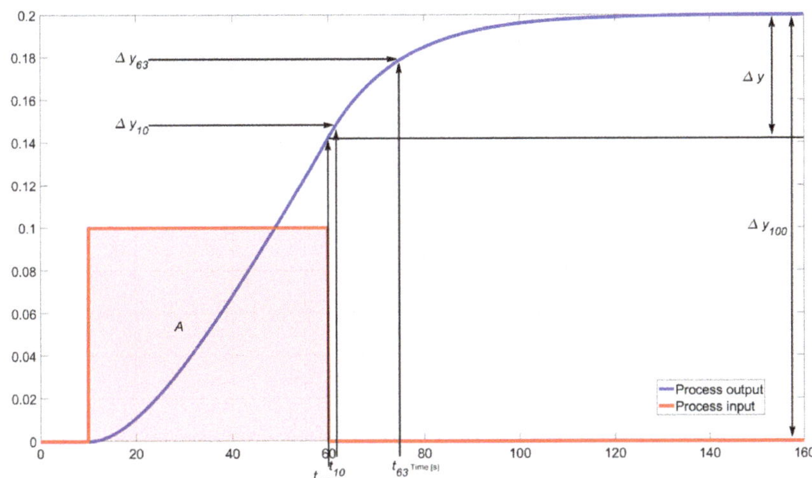

Figure 5. Process identification from wide pulse response for systems with transport delay (wide pulse process excitation—red line, process output—blue line).

- The time delay of the IFOPDT model τ can be obtained by measuring the time required for the process response to reach 10% of Δy, marked as t_{10}. Therefore, the delay is calculated as $\tau = t_{10} - t_{uOFF}$, wherein t_{uOFF} is the moment when the excitation ends;
- The equivalent time constant T is the time for the response to change from 10% to 63% of Δy, $T = t_{63} - t_{10}$;
- Process gain K can be determined as mentioned as the slope of the reaction curve (system output), or more robustly as the ratio of a total change of output and area of wide pulse excitation A, $K = \Delta y_{100} / A$.

3.2. Process with Moderate or Significant Transport Delay

If the process has a transport delay that cannot be neglected (as shown in Figure 6), the IFOPDT model parameters can be obtained by the following procedure:

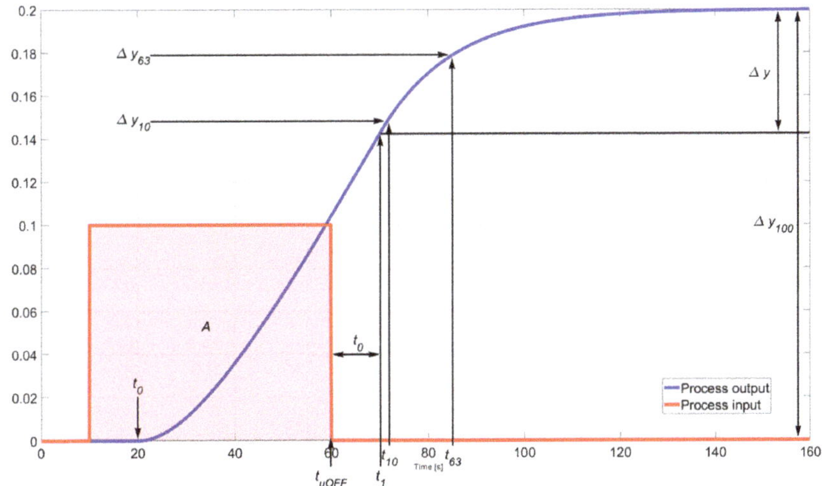

Figure 6. Process identification from wide pulse response for systems with transport delay.

- The time delay of the IFOPDT model τ can be obtained as a sum of the delay t_0 and time $t_{10} - t_1$. Time t_1 is defined as the sum of wide pulse end time t_{uOFF} and t_0, $t_1 = t_{uOFF} + t_0$. Time t_{10} is calculated by measuring the time required for the process response to reach 10% of Δy;
- The equivalent time constant T is the time for the response to change from 10% to 63% of Δy, $T = t_{63} - t_{10}$;
- Process gain K can be determined as mentioned as the slope of the reaction curve (system output), or more robustly as the ratio of a total change of output and area of wide pulse excitation A, $K = \Delta y_{100} / A$.

3.3. PID Tuning Rules

For processes with the integrator, a PD controller should be used. Due to load rejection and setting nominal regime, the integral part will also be included, but integral time should have a high value according to process time constant T.

The structure of the PID controller is defined as:

$$G_{PID} = K_p \left(1 + \frac{1}{T_i s} + \frac{T_d s}{T_f s + 1} \right). \tag{1}$$

First, the controller will be observed as simple PD, $G_{PD}(s) = K_p(1 + T_d s)$. After determining K_p and T_d, filtering of differential action T_f, the integral term of the PID controller T_i will be introduced and it will be assumed that they do not have a lot of influence on stability and system performance. By introducing the integral action, the order of astatism of the system increases, and the stability of the closed-loop system, generally speaking, may be impaired. With certain approximations, it can be shown that by introducing the integral effect, the transfer function in the open-loop discrete system gets a pole at the point $p = 0$ and zero at the point $z = -1/T_i$. By choosing a sufficiently large T_i parameter e.q. $T_i \geq 10T$, it is possible for this pole and zero to be close enough to each other, but also far enough from the point ω_1 (frequency of phase margin) so that their influence on the phase margin, and consequently the stability, becomes almost negligible.

The steps for determining the PD parameters are based on [12,14]. The differential time constant T_d is taken to be

$$T_d = T, \tag{2}$$

where T is the equivalent time constant of the process, evaluated as shown in Sections 3.1 and 3.2. Then open-loop transfer function is:

$$W(s) = G(s)G_{PD}(s) = \frac{K}{s(Ts + 1)} e^{-\tau s} K_p (1 + T_d s), \tag{3}$$

$$W(s) = \frac{KK_p}{s} e^{-\tau s}. \tag{4}$$

The pole of the process $G(s)$ is canceled by proportional and differential action of the PD controller $(1 + T_d s)$. The proportional gain of the PD controller K_p remains to be determined, which can be determined by applying the Nyquist stability criterion to the characteristic equation

$$1 + W(s) = 1 + G(s)G_{PD}(s) = 1 + \frac{KK_p}{s} e^{-\tau s} = 0. \tag{5}$$

Choosing KK_p to obtain the desired phase margin ϕ_{pm} [14,18,19] gives

$$K_p = \frac{\mu}{K\tau} \tag{6}$$

$$\mu = \frac{\pi}{2} - \phi_{pm} \tag{7}$$

where ϕ_{pm} is the desired phase margin [16,17].

Parameter μ is chosen to be in the range $\mu \in (0.32, 0.54)$. A higher value of μ results in better performance, more aggressive control, and less robustness.

The integral term of the PID controller T_i is introduced in the controller for good load rejection and regime changing. It should have a significantly higher value than the process dynamic but small enough to preserve good disturbance rejection [19], and the adoption of the value $T_i = 10T$ is suggested.

To avoid aggressive control caused by a high gain of differential action T_d, a filter for a differential part is used, with the time constant $T_{fd} = T/n_d$. Parameter n_d is chosen from the range $n_d \in (5, 10)$. Lower n_d values are better for good filtration of measurement noise and higher values for stability because the filter has less influence on the closed-loop process. It is a heuristic recommendation, and it is formed in such a way as to make a compromise between noise-measurement filtering that can cause significant damage when realizing the differential effect of the PID controller and preserving the system's dynamics. Paper [20] analyzed various filters and parameters, and the authors of this paper chose the proposed one due to the implementation simplicity and the intuitiveness of the setting parameter. It is up to the system designer to find a compromise between these two conflicting requirements. The recommendation is that n_d be between 5 (when it is up to the dynamics of the system) and 10 (when the measurement signal filtration is more important) as the range that the authors consider satisfactory, after many experiments with different systems. The recommended value of $n_d = 10$ can be adopted during implementation, and if the control is too aggressive, the value of n_d can be reduced to 5.

3.4. Simulation Results of the Proposed Tuning Procedure

As an example, for comparison with IMC-PID [21], $G_p = \frac{0.004}{s(15s+1)}e^{-\tau s}$ will be used. IMC-PID (internal model control–PID) is the controller-tuning procedure known in the literature [22] and frequently used in industrial practice. The author's idea was to compare the result of the proposed controller with the result of this well-known tuning procedure.

First, it is necessary to model the process and identify parameters of the proposed IFOPTD model. In $t = 10$ s wide pulse excitation is started (Figure 7). From the process response, it can be seen that the process had no significant time delay. In this example, the duration of the wide pulse excitation was $\Delta t = 100$ s. At $t_{uOFF} = 110$ s, pulse excitation ended. Then, the time constant and the transport delay could be calculated as in the proposed procedure.

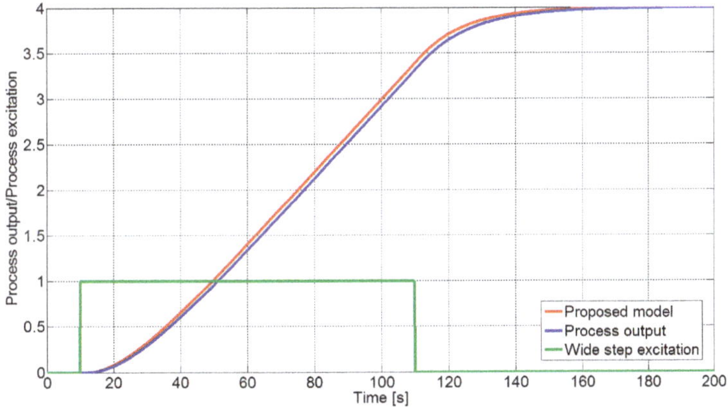

Figure 7. Wide step response of the process (blue) and proposed IFOPDT model (red), wide step excitation (green) scaled 10 times due to representation (magnitude of wide step excitation is $u = 10$).

Process gain is estimated as the ratio of the process output change and the area of wide pulse excitation on the process input $K = \Delta y_{100}/A$. Area A is defined as a product of wide pulse excitation duration $\Delta t = 100s$ and excitation magnitude $u = 10$, so $A = 1000$. The final value of the process output is $\Delta y_{100} = 4$. The estimated parameters of G_p are as follows:

$$K = 0.004, T = 13.5, \tau = 1.8 \tag{8}$$

Figure 7 shows promising results of the proposed IFOPDT modeling for a process chosen in [21]. The main goal for the controller is to satisfy two opposite demands: to have high robustness and good performance, so for analysis, parameter μ was chosen to be $\mu = 0.32$ and $\mu = 0.54$, for good robustness and good performance, respectively. Parameter T_d was chosen to be identical to the estimated time constant, $T_d = T$. For filtering differential action, the time constant of the filter $T_{fd} = T_d/5$ was chosen. It was high enough to suppress the impact on the PD output but at the same time small enough not to affect the process's dynamic. Reasons for introducing an integral part in the controller have been mentioned. The time constant of integral action T_i was chosen to be much higher than the process dynamics, $T_i = 10T$. Thus, the parameters of the PID controller were

$$K_p = 44.4 \ (\mu = 0.32), \ K_p = 75.0 \ (\mu = 0.54), \ T_d = 13.5, \ T_i = 135, \ T_{fd} = 2.25. \tag{9}$$

Figure 8 shows a comparison between the proposed procedure for two values of tuning parameter μ and the tuning presented in [21], also suggested for integrating processes and classical Ziegler Nichols tuning, based on a relay experiment [23].

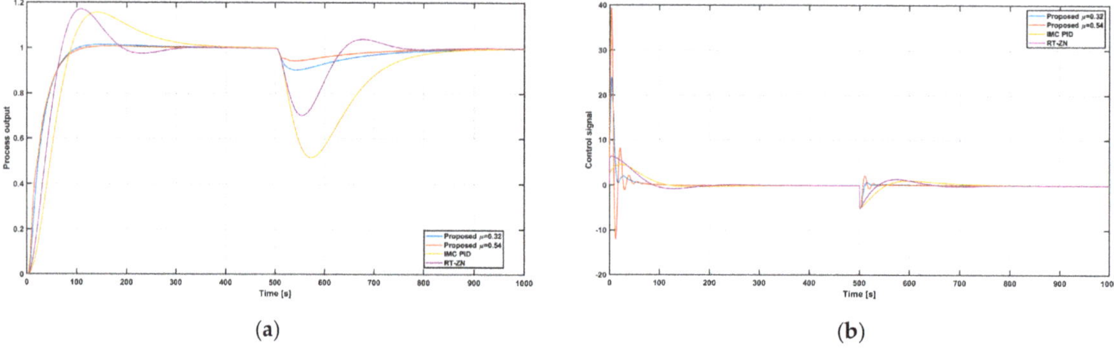

Figure 8. Comparison of PID controllers: proposed "robust" PID for $\mu = 0.32$ (blue), proposed "performance" PID for $\mu = 0.54$ (blue), IMC PID (violet), and RT-ZN PID (yellow). (**a**) Process output, (**b**) control signal.

A disturbance is introduced as a step signal on the process input in the middle of simulation, $t_{dist} = 500$ s. Good disturbance rejection of the proposed tuning method can be seen in Figure 8, as a result of a good selection of the integral time, $T_i = 10T$.

4. Experimental Results and Verification

To carry out the experimental verification of the proposed method based on wide pulse response, preparation for experimental tuning was carried out. Namely, two methods were selected for comparison, the proposed WPRT and the classic Ziegler Nichols method, with the identification of process parameters using the relay-tuning procedure. In the case of the water-level control in the steam drum (separator), the control structure for the two experiments is shown in Figure 3, in such a way that by selecting the operating mode, one chooses the control via the PID controller (mode 1), relay experiment (mode 2), and wide pulse (mode 3).

1. Mode 1—PID operating mode.
2. Mode 2—mode for obtaining the tuning parameters by the ZN method, so-called relay tuning (RT-ZN). In this mode, an error signal is fed to the input of the relay, which is formed as the difference between the setpoint level and the level measured in the separator, and the control signal is generated at the output of the relay, which is forwarded to the flow regulators of the pumps. In this way, controlled self-oscillations are obtained in the system, the basis for determining the parameters K_u and T_u for the PID controller tuned by the ZN method. To make the method reliable for identifying the amplitude and period of the oscillations, it is necessary to wait for the formation of self-oscillations and their duration of at least five periods. This requires a certain amount of time necessary for careful monitoring of the experiment. If the output value of the level exceeds some values set as dangerous for the process, it is required to take over manual control and bring the process into the safe operating zone.
3. Mode 3—mode for applying the WPRT method implies breaking the feedback loop, using a constant control signal from the last period of automatic operation, setting a step excitation signal, and waiting for the appropriate time. Then the excitation returns to the previous level. After the end of the transition regime, as described in the previous section, the process returns to the automatic operation mode.

The obtained measurements and time response for the proposed WPRT method are shown in Figure 9. The system was switched to manual mode at $t = 95$ s, with the feedwater flow per pump set at $u = 138$ kg/s. A pulse was set at $t_{uON} = 100$ s with an amplitude $\Delta u = 30$ kg/s and duration $t_{pulse} = 20$ s. The water level in the steam separator ranged from 5.5 m to 7.3 m, which was within acceptable limits for reliable operation. The automatic operation mode was set again at $t = 150$ s. Based on this experiment, the IFOPDT model parameters were determined as suggested in the previous section, and then the PID controller parameters were determined. As it is necessary to achieve maximum performance in this subsystem, the parameter $\mu = 0.54$ was selected, and the following PID controller parameters were obtained

$$K_p = 36.7, \; T_i = 35, \; T_d = 3.5, \; T_{fd} = 0.7 \tag{10}$$

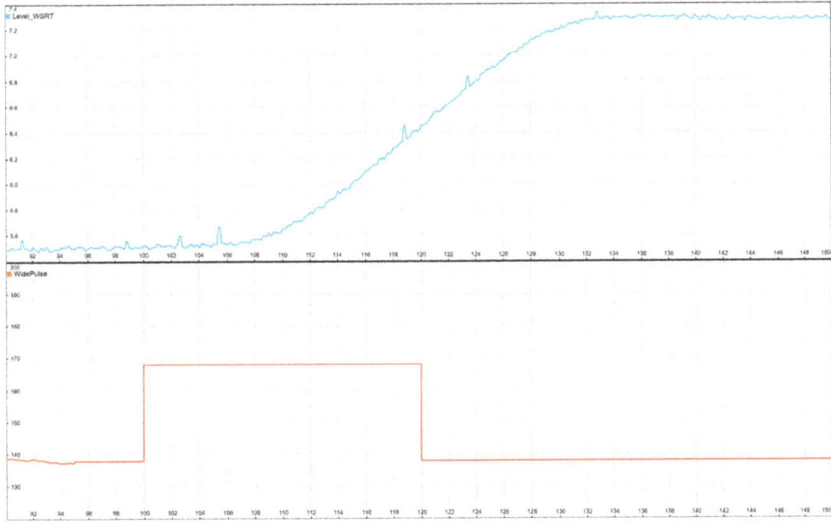

Figure 9. Wide step response of process (blue), wide step excitation (red): flow setpoint (magnitude of wide step excitation is $u = 168 - 138 = 30$ kg/s), duration of pulse excitation is 20 s, end of the experiment $t_{end} = 150$ s.

After PID controller tuning with two methods, RT-ZN and WPRT, the following comparative experiments were performed, at the same nominal power of the plant and in the quiet operation of the block, without major disturbances:

4. Reference change test: The water level setpoint in the separator was changed from 5.5 m to 6.5 m.
5. Process input disturbance test: The output of the controller, the flow rate of the feed water for the slave controller, was increased by 10 kg/s.
6. Process output disturbance test: At the output of the process, the level measurement itself was increased by 0.3 m.
7. Controller nominal operation test: In the nominal operation mode, the control deviation was recorded to compare the two settings of the controller.

All tests lasted 250 s and were repeated four times with both controller settings, with alternating repetitions, since there are always minor disturbances in the process caused by uneven coal quality, uneven combustion, and many other parameters. To assess the control quality, the integral absolute criterion was applied, $IAE = \int |e(t)|\, dt$, which had the control deviation as an input $e(t) = r(t) - y(t)$. The control signals had to be realistic since there were slope limiters in the system that did not allow the control signal to rise or fall too much. It should be noted that all signals were recorded by the primary sampling time that the DCS system itself operates with, $T_s = 0.2$ s.

The following Figures 10–13 illustrate time diagrams for four experiments for each controller. They can be used to compare the time responses and the control quality for the classic RT-ZN method, and the proposed WPRT method in actual application (real conditions) in an extremely complex process being controlled.

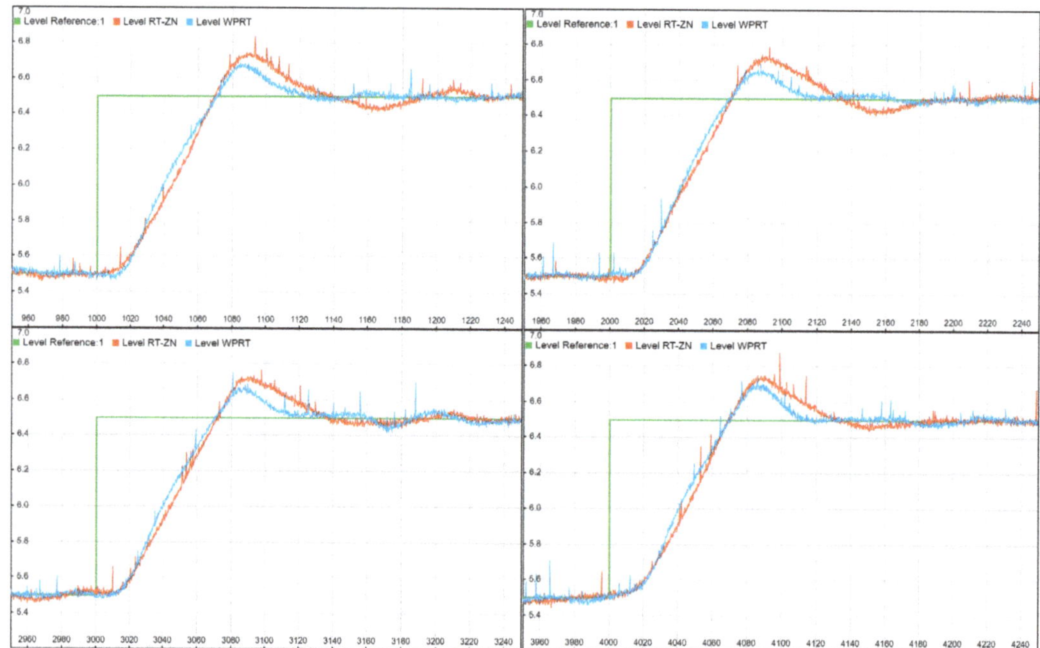

Figure 10. Setpoint change experiment: The water-level setpoint in the separator was changed from 5.5 m to 6.5 m. Water-level setpoint (green), RT-ZN tuning (red), proposed WPRT method (blue).

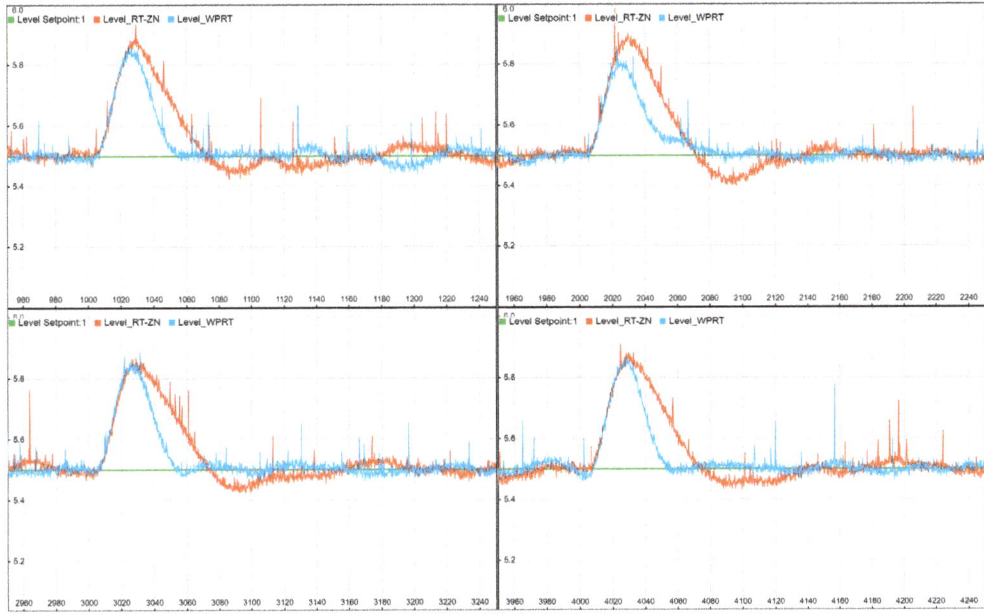

Figure 11. Process input disturbance test: The controller's output, the setpoint flow rate of the feed water for the slave controller, was increased by 10 kg/s. Water-level setpoint (green), RT-ZN tuning (red), proposed WPRT method (blue).

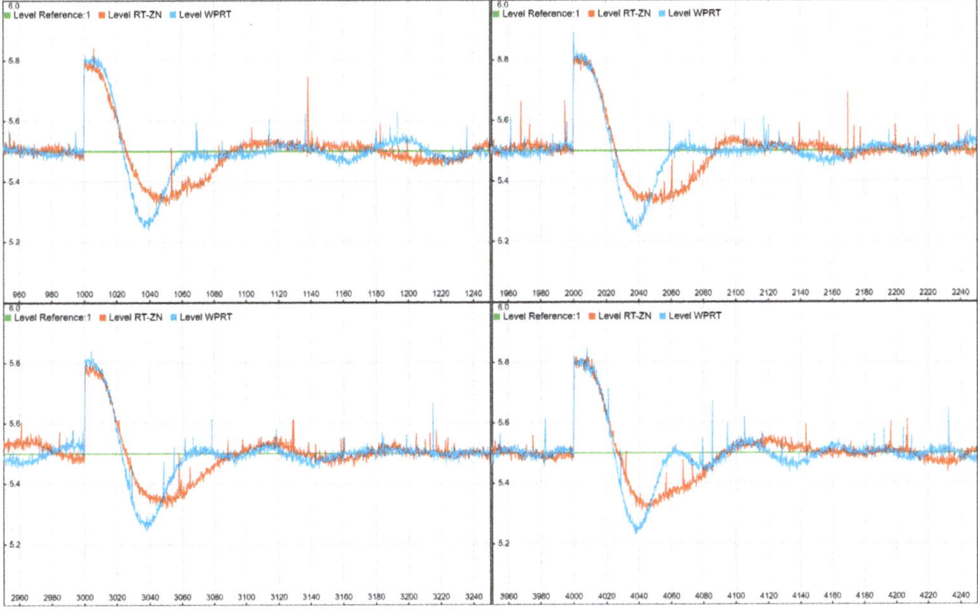

Figure 12. Process output disturbance test: At the process's output, the level measurement was artificially increased by 0.3 m. Water-level setpoint (green), RT-ZN tuning (red), proposed WPRT method (blue).

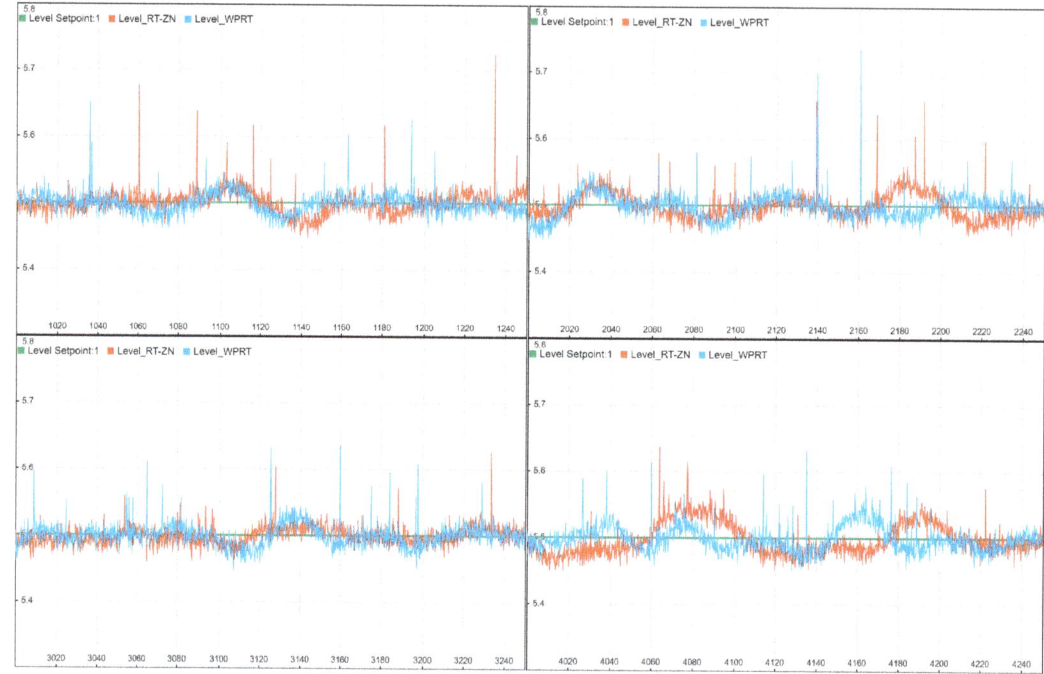

Figure 13. Controller nominal operation test: In the nominal operation mode, the deviation was recorded to compare the two tunings of the controller. Water-level setpoint (green), RT-ZN tuning (red), proposed WPRT method (blue).

5. Discussion

After thoroughly conducting experiments in an existing plant facility, it is possible to perform a results-based analysis. Table 1 shows in detail the values of the IAE criteria for each type and for each experiment repetition. The last row shows the mean values for each experiment for both methods.

Table 1. Table of comparative results of integral absolute error (IAE) for four different experiments, each repeated four times.

Test Type	Setpoint Change IAE		Input Disturbance IAE		Output Disturbance IAE		Nominal Regime IAE	
Test No	WPRT	RT-ZN	WPRT	RT-ZN	WPRT	RT-ZN	WPRT	RT-ZN
1	47.44	54.80	11.82	17.18	13.96	14.64	2.90	3.21
2	46.78	54.52	10.78	17.82	13.31	14.75	3.57	3.85
3	48.15	53.69	11.07	17.02	12.78	13.52	3.18	2.77
4	47.69	53.02	10.89	17.06	14.26	15.28	3.80	5.04
Average	47.52	54.01	11.14	17.27	13.58	14.55	3.36	3.72

The lower value of the IAE criteria for the proposed WPRT method is evident compared to the tuning based on the relay-tuning experiment (RT-ZN), which is common in practice.

Figure 14 shows a comparative illustration of the IAE criteria mean values for individual experiments. Lower criteria values reflect a better behavior of the controller in operation, and the consistency of improvement for all proposed tests should be emphasized. Thus, the better performance of the proposed WPRT method is unequivocally demonstrated. This claim is supported by the results obtained during the theoretical analysis of the control quality of several controllers, as shown in Figure 8. Since there were no options for a large number of experiments to be conducted in practical conditions, the comparison was made only with the RT-ZN method, the method often used in practice, which showed promising results in the previous analysis. It should be noted that the control signals are not shown because they were limited by rate limiters located before the actuators (Voith's hydraulic couplings). Therefore, they did not affect the analyzed results.

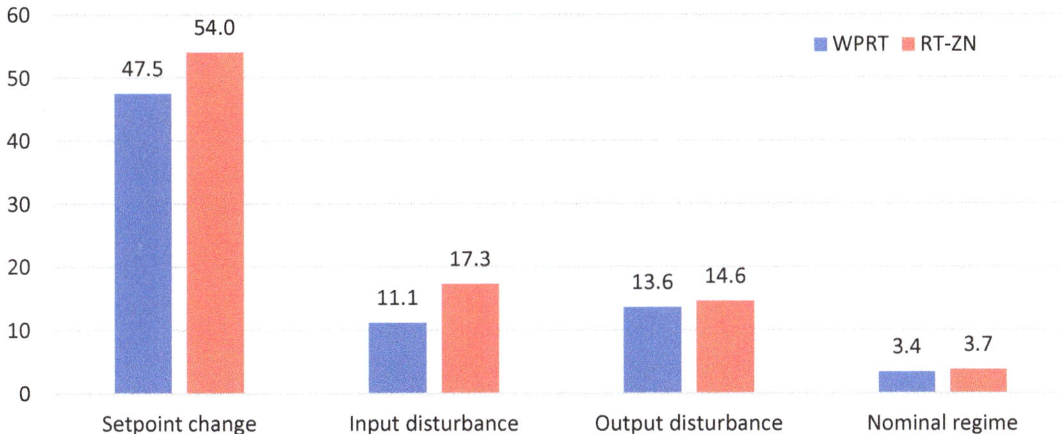

Figure 14. The mean integral absolute error (IAE) values for the performed tests, RT-ZN tuning (red), proposed WPRT method (blue).

For future research, signal-processing methods and measurement filtering should be considered. As visible from the signals from the plant, there was an evident level of noise that affected the quality of parameter estimation (Figure 9), and the particularly problematic issue is that this noise did not have Gaussian distribution. Namely, bubbles often developed in the steam separator in water, traveling to the surface and being detected by the level sensor as peaks. It is necessary to carry out high-quality data pre-processing but in such a way as not to affect the dynamics of the signal, because it could affect the estimation of the IFOPDT model parameters. Recently, many deep learning-based methods [24–27] have been developed, so it is necessary to perform an analysis and try to apply them in industrial practice.

6. Conclusions

The paper presents a novel method for the PID controller parameter tuning for control processes with the integrator. This type of process is complicated and challenging to control because instability and poor performance are typical with controllers tuned using conventional tuning procedures. Within the proposed WPRT, clear instructions have been provided for conducting a quick, simple-to-understand, and easily controlled experiment to obtain the IFOPDT model parameters. After identifying the parameters of the model, tuning of the PID controller is performed, with one free tunable parameter available, which can be used to choose between better performance or robustness.

During the actual process of the water-level control in the steam separator at the thermal power plant, the controller tuning procedure is demonstrated. A comparative analysis with a classically tuned controller using a relay experiment is provided. The results of the plant operation show significantly better results than the method proposed herein.

The proposed WPRT method can be effectively applied to a wide range of industrial processes that include the integrator, usually level controllers, and positional servo systems. The proposed method proved to be better in all aspects of the application: from the simple experiment to the behavior of the controlled output value, but also due to the possibility of additional tuning by using a free parameter. The authors believe that this paper represents the missing segment in the range of various techniques for PID controller tuning, especially for designing and controlling industrial facilities, such as the water-supply system in thermal energy blocks.

Author Contributions: Conceptualization: G.S.K. and Z.M.D.; methodology, software, validation, formal analysis, investigation, resources, data curation, writing—original draft preparation, visualization: G.S.K.; writing—review and editing, supervision, project administration, funding acquisition: Z.M.D. All authors have read and agreed to the published version of the manuscript.

Funding: This research was funded by the Ministry of Education, Science and Technological Development of the Republic of Serbia, grant number 451-03-68/2022-14/200103.

Data Availability Statement: All the data are in the manuscript.

Conflicts of Interest: The authors declare no conflict of interest.

References

1. Yamamoto, S.; Hashimoto, I. Resent status and future needs: The view from Japanese industry. In Proceedings of the Fourth International Conference on Chemical Process Control, Padre Island, TX, USA, 17–22 February 1991.
2. Hansen, L.; Bram, M.V.; Pedersen, S.; Yang, Z. Performance Comparison of Control Strategies for Plant-Wide Produced Water Treatment. *Energies* **2022**, *15*, 418. [CrossRef]
3. Durdevic, P.; Yang, Z. Application of H∞ Robust Control on a Scaled Offshore Oil and Gas De-Oiling Facility. *Energies* **2018**, *11*, 287. [CrossRef]
4. Hansen, L.; Durdevic, P.; Kasper; Jepsen, L.; Yang, Z. Plant-wide Optimal Control of an Offshore De-oiling Process Using MPC Technique. *IFAC-PapersOnLine* **2018**, *51*, 144–150. [CrossRef]
5. Hofreiter, M. Relay Identification Using Shifting Method for PID Controller Tuning. *Energies* **2021**, *14*, 5945. [CrossRef]
6. Rotac, V.J. *Rascet Nastrojki Promyšlenych Sistem Regulirovanija*; Gosenergoizdat: Moskva, Russia, 1961.
7. Åström, K.J.; Hägglund, T. Automatic tuning of simple regulators with specifications on phase and amplitude margins. *Automatica* **1984**, *20*, 645–651. [CrossRef]
8. Hang, C.C.; Astrom, K.J.; Wang, Q.G. Relay feedback auto-tuning of process controllers—A tutorial review. *J. Process Control.* **2002**, *12*, 143–162. [CrossRef]
9. Shen, S.-H.; Wu, J.-S.; Yu, C.-C. Use of biased-relay feedback for system identification. *AIChE J.* **1996**, *42*, 1174–1180. [CrossRef]
10. Li, W.; Eskinat, E.; William, L.L. An improved autotune identification method. *Ind. Eng. Chem. Res.* **1991**, *30*, 1530–1541. [CrossRef]
11. Zeng, D.; Zheng, Y.; Luo, W.; Hu, Y.; Cui, Q.; Li, Q.; Peng, C. Research on Improved Auto-Tuning of a PID Controller Based on Phase Angle Margin. *Energies* **2019**, *12*, 1704. [CrossRef]
12. Kvascev, G.; Djurovic, Z.; Vlatkovic, V. A Wide Pulse Response Procedure for Tuning of PD/PID Controller for Integrating Processes. In Proceedings of the 2015 International Conference on Computer and Computational Sciences (ICCCS), IEEE, Greater Noida, India, 27–29 January 2015; pp. 54–58.
13. O'Dwyer, A. *Handbook of PI and PID Controller Tuning Rules*; Imperial College Press: London, UK; World Scientific: Singapore, 2009; ISBN 978-1-84816-242-6. [CrossRef]
14. Matausek, M.R.; Kvascev, G.S. A unified step response procedure for autotuning of PI controller and Smith predictor for stable processes. *J. Process Control.* **2003**, *13*, 787–800. [CrossRef]
15. Brkić, L.; Živanović, T. *Steam Boilers*; Mašinski fakultet Univerziteta u Beogradu: Beograd, Serbian, 2007. (In Serbian)
16. Mićević, Z. *Boiler Installations*; Energoprojekt Enel d.d.: Beograd, Serbian, 1999. (In Serbian)
17. Flynn, D. *Thermal Power Plant Simulation and Control*; The Institution of Electrical Engineering: London, UK, 2003.
18. Matausek, M.R.; Micic, A.D. A modified Smith predictor for controlling a process with an integrator and long dead-time. *IEEE Trans. Autom. Control.* **1996**, *41*, 1199–1203. [CrossRef]
19. Matausek, R.; Micic, A.D. On the modified Smith predictor for controlling a process with an integrator and long dead time. *IEEE Trans. Autom. Control.* **1999**, *44*, 1603–1606. [CrossRef]
20. Hägglund, T. Signal Filtering in PID Control. *IFAC Proc. Vol.* **2012**, *45*, 1–10. [CrossRef]
21. Arbogast, J.E.; Cooper, D.J. Graphical Technique for Modeling Integrating (Non-Self-Regulating) Processes without Steady-State Process Data. *Chem. Eng. Commun.* **2007**, *194*, 1566–1578. [CrossRef]
22. Rivera, D.E.; Morari, M.; Skogestad, S. Internal model control: PID controller design. *Ind. Eng. Chem. Process Des. Dev.* **1986**, *25*, 252–265. [CrossRef]
23. Ziegler, J.G.; Nichols, N.B. Optimum Settings for Automatic Controllers. *Trans. ASME* **1942**, *64*, 759–768. [CrossRef]

24. Djordjevic, N.; Dzamic, N.; Stojic, A.; Kvascev, G. Denoising the open-loop step response using an encoder-decoder convolutional neural network. In Proceedings of the 9th International Conference on Electrical, Electronic and Computing Engineering IcETRAN 2022, Novi Pazar, Serbia, 6–9 June 2022.
25. Antczak, K. Deep recurrent neural networks for ECG signal denoising. *arXiv* **2018**, arXiv:1807.11551.
26. Zhu, W.; Mousavi, M.; Beroza, G. Seismic signal denoising and decomposition using deep neural networks. *IEEE Trans. Geosci. Remote Sens.* **2019**, *57*, 9476–9488. [CrossRef]
27. Li, X.; Liu, J.; Li, J.; Li, X.; Yan, P.; Yu, D. A Stacked Denoising Sparse Autoencoder Based Fault Early Warning Method for Feedwater Heater Performance Degradation. *Energies* **2020**, *13*, 6061. [CrossRef]

Review

Direct Contact Condensers: A Comprehensive Review of Experimental and Numerical Investigations on Direct-Contact Condensation

Paweł Madejski *, Tomasz Kuś, Piotr Michalak, Michał Karch and Navaneethan Subramanian

Department of Power Systems and Environmental Protection Facilities, Faculty of Mechanical Engineering and Robotics, AGH University of Science and Technology, Al. Mickiewicza 30, 30-059 Krakow, Poland
* Correspondence: madejski@agh.edu.pl

Abstract: Direct contact heat exchangers can be smaller, cheaper, and have simpler construction than the surface, shell, or tube heat exchangers of the same capacity and can operate in evaporation or condensation modes. For these reasons, they have many practical applications, such as water desalination, heat exchangers in power plants, or chemical engineering devices. This paper presents a comprehensive review of experimental and numerical activities focused on the research about direct condensation processes and testing direct contact condensers on the laboratory scale. Computational Fluid Dynamics (CFD) methods and CFD solvers are the most popular tools in the numerical analysis of direct contact condensers because of the phenomenon's complexity as multiphase turbulent flow with heat transfer and phase change. The presented and developed numerical models must be carefully calibrated and physically validated by experimental results. Results of the experimental campaign in the laboratory scale with the test rig and properly designed measuring apparatus can give detailed qualitative and quantitative results about direct contact condensation processes. In this case, the combination of these two approaches, numerical and experimental investigation, is the comprehensive method to deeply understand the direct contact condensation process.

Keywords: direct contact heat exchanger; direct contact condensation; CFD modeling; test rig

Citation: Madejski, P.; Kuś, T.; Michalak, P.; Karch, M.; Subramanian, N. Direct Contact Condensers: A Comprehensive Review of Experimental and Numerical Investigations on Direct-Contact Condensation. *Energies* **2022**, *15*, 9312. https://doi.org/10.3390/en15249312

Academic Editor: Ahmed Abu-Siada

Received: 10 October 2022
Accepted: 4 December 2022
Published: 8 December 2022

Publisher's Note: MDPI stays neutral with regard to jurisdictional claims in published maps and institutional affiliations.

1. Introduction

Direct Contact Condensers (DCCs) have been used in industry since the beginning of the 20th century [1], covering a wide range of various applications in chemical engineering, water desalination, air conditioning, and energy conversion processes.

In this device, the cooling liquid is directly mixed with gas or vapour, which results in condensation and a significant decrease in device volume [2]. Involving a surface condenser of the same capacity direct condenser has several advantages. Due to direct contact with process fluids, its construction is simpler and more corrosion resistant [3], less expensive [4], easier to maintain, and simpler in operation [5].

Direct contact condensers are generally divided into spray-type, film-type, and bubbling type [6]. In the first solution, the sprayed liquid phase flows downwards and is in contact with flowing upwards gas. In the second case, both phases flow counter currently. In the latter solution, the bubbling gas phase passes through the liquid layer. Furthermore, spray condensers can exist with constant pressure or constant area jet ejectors [7]. Despite these apparatuses' wide range of applications, plenty of studies summarize theoretical and practical aspects of their development. Aidoun et al. [8,9] presented results of experimental and numerical studies focusing on ejectors and their applications in refrigeration systems. Mil'man and Anan'ev [10] focused on the application of air-cooled condensing units in thermal power plants. Xu et al. [11] discussed recent advances in humidification-dehumidification desalination processes, including direct and indirect contact condensers. They are also

     59

commonly used as dehumidifiers in solar-driven humidification-dehumidification desalination [12,13] and seawater greenhouse [14] systems. The application of direct and indirect condensers in the pyrolysis of waste plastics was discussed by Kartik et al. [15], and in pyrolysis of biomass to bio-oil was presented in [16].

The wide range of possible and actual applications of these devices indicates the need to investigate more deeply the direct contact condensation process by using simulation and experimental studies devoted to developing direct contact condensers. This paper aims to present various analyses and their main outcomes to give the full view of the most important factors that must be considered during theoretical and experimental research.

2. Direct Contact Condensers

Direct Contact Heat Exchangers (DCHEs) play an important role in various technological processes, including humidifying air, cooling water, and removing excess heat from flue gases. The exchange processes in such apparatuses occur under contact with the liquid phase (e.g., water) and gas phase (e.g., air) at the interphase surface. In this case, heat and mass transfer are mainly determined by the geometric dimensions of the surface area for contact between the two phases. A specific value of this surface area (attributed, for instance, to the volume of the active zone of DCHE) depends on the method of interaction of the contacting phases, i.e., on the DCHE design. The most commonly used designs of heat exchangers are the following [6]:

- spray-type (gas phase flows upwards and comes into contact with the liquid phase, which is sprayed from the nozzles and flows downwards),
- film-type (liquid phase flows downwards as a thin liquid film on the inside wall of the vertical tube while the air flows counter currently) [17–20],
- bubbling-type (bubbling of the gas phase through a layer of the liquid located on a hole tray [21–23] or in a vertical channel [24,25].

Direct Contact Condensers (DCCs) have a variety of purposes. They can be used to heat the liquid for heat recovery. The hot liquid can be used to heat rooms, preheat raw materials, or melt solids such as ice. DCCs can be used to cool the gas to generate condensate. Condensate can be used to purge a reaction product, such as acids coabsorbed in the DCC, condense a particulate by converting it from a vapor to a liquid or solid, and grow particulate by condensing directly on the particulate surface to improve its capture or reclaim water in arid regions. Direct Contact Condensers also can reduce gas volume, suppress stack plumes, and lower energy requirements.

2.1. Type of Direct Contact Condensers

In direct-contact condensers, the gas and liquid come in direct contact. The cooling liquid is sprayed into the gas region to start a rapid condensation, which maximizes the thermal efficiency of condensers. The heat is transferred from a gas to a liquid, and the condensate temperature is the same as that of the cooling liquid leaving the condenser. The condensate cannot be reused as feed water if the cooling water is not pure and free from harmful impurities. The occurrence of the other gases strongly impacts the heat transfer rate and condensation efficiency. This process is one of the important issues investigated experimentally or numerically to determine overall efficiency and properly design Direct Contact Condensers. The general classification of condensers is presented in Figure 1.

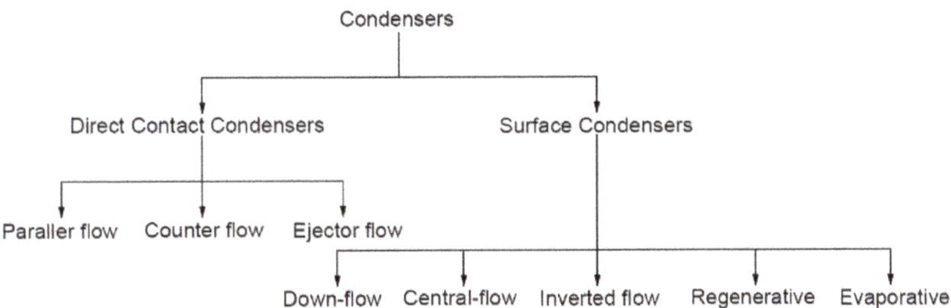

Figure 1. Classification of condensers.

In a parallel flow jet type condenser, the exhaust steam and cooling water find their entry at the top of the condenser and then flow downwards, and condensate and water are finally collected at the bottom (Figure 2).

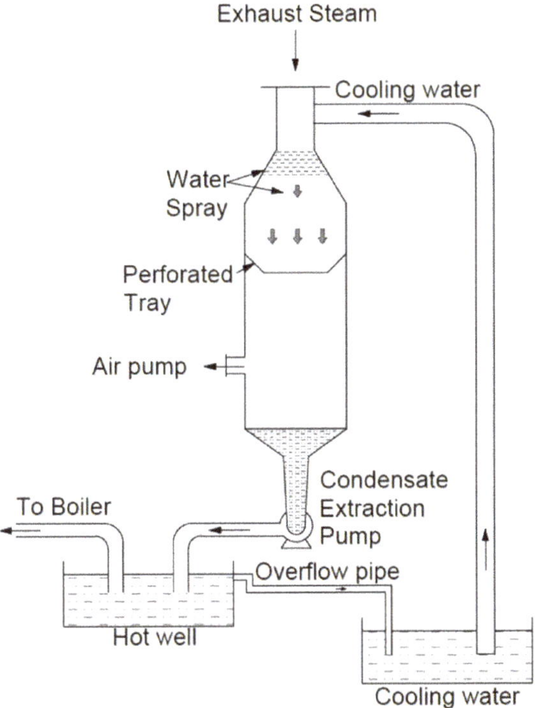

Figure 2. Parallel flow type condenser.

The steam and cooling water enter the condenser from opposite directions in a counter jet type condenser. Generally, the exhaust steam travels upward and meets the cooling water, which flows downwards. In this low-level jet-type condenser (counter jet type condenser), presented in Figure 3, the exhaust steam enters slightly lower than in a parallel flow jet-type condenser, and the cooling water is supplied from the top of the condenser chamber (Figure 3). The direction of the steam is upward, and the cooling water is downward. An air pump creates a vacuum and is placed on top of the condenser. The vacuum sucks the cooling water, and a hollow cone plate collects the falling water, which joins the second series of streams and meets the exhaust steam entering from below. The resulting condensate is delivered to the tank through a vertical pipe by the condensate pump. Another solution of counter jet type condensers is called barometric condenser and is presented in Figure 4. In this type, the shell is placed at the height of about 10.363 m above the hot well; thus, there is no need to provide an extraction pump. Provision of providing injection pump is observed, where water under pressure is unavailable.

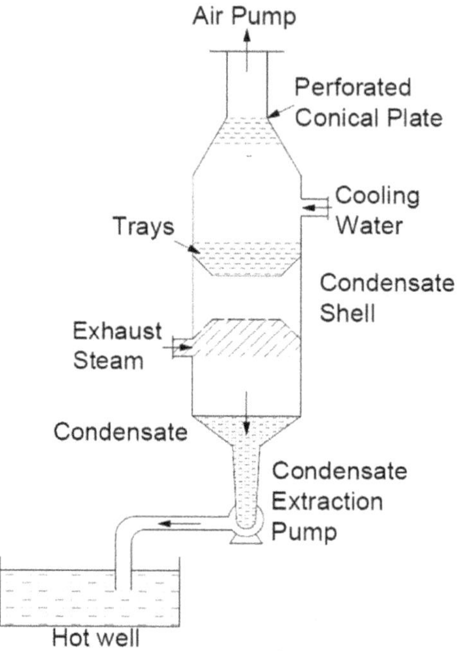

Figure 3. Low-level counter-flow jet type condenser.

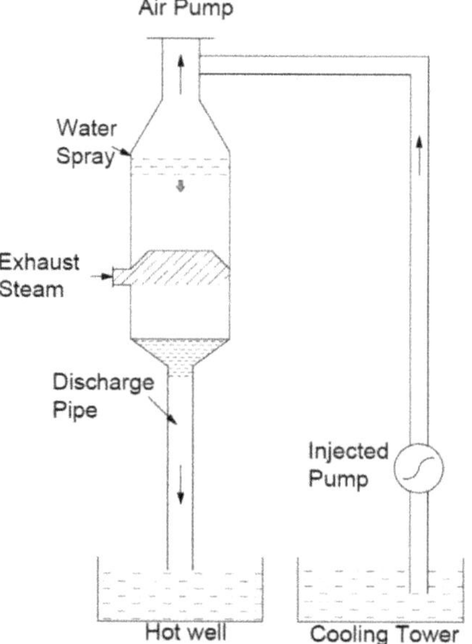

Figure 4. High-level counter-flow jet type condenser.

In Figure 4, the discharge pipe is connected to the bottom of the condenser shell. The exhaust steam enters the system in the lower part of the condenser, with the flow direction pointing upwards. Cooling water enters at the top and is collected by a punched cone plate. An air pump creates the vacuum on top of the shell. Steam and cooling water mix together and are carried through a discharge pipe to the tank. The difference between low and high-level jet condensers is that there is no pump between the tank and the discharge pipe in the high-level type.

The last type of jet condenser is an ejector flow jet type condenser (Figure 5). Here the exhaust steam and cooling water mix in hollow truncated cones. Due to this decreased pressure, exhaust steam and associated air are drawn through the truncated cones, finally leading to the diverging cone. In the diverging cone, a portion of kinetic energy is converted into pressure energy which is more than the atmospheric, so that condensate consisting of condensed steam, cooling water, and the air is discharged into the hot well. The exhaust steam inlet is provided with a non-return valve which does not allow the water from the hot well to rush back to the engine in case of cooling water supply to the condenser.

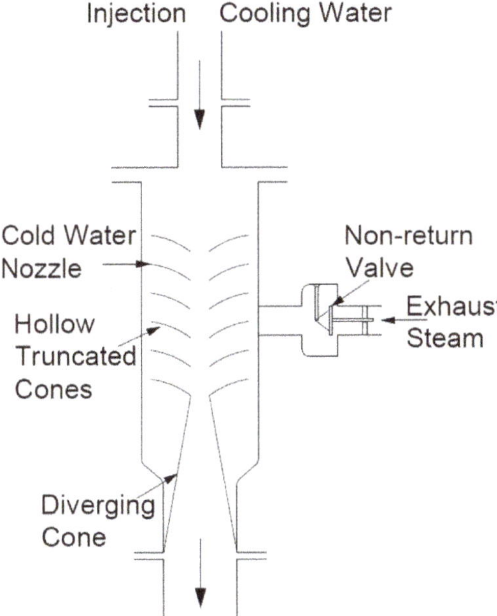

Figure 5. Ejector-flow jet type condenser.

The cooling cycle makes use of a steam ejector condenser. The steam ejector condenser is classified into two types based on the mixing method in the primary nozzle exit [7,26]. The first one is the constant pressure jet ejector (CPJE), and the other one is the constant area jet ejector (CAJE). The CPJE performs better than the CAJE due to better turbulent mixing [7,27]. In addition to having no moving parts, the steam ejector condenser benefits from lower maintenance and capital cost than the compressor.

2.2. Water and CO_2 Separation

After the exhaust passes through the Direct Contact Condenser (DCC) for condensation, the condensate water from the DCC consists of a proportion of non-condensable gases such as CO_2, air, or other gases. The stream is passed through the separator or non-condensable gas removal system, which separates the water and the non-condensable gases. By this method, the separated CO_2 can be sent to the CCU unit for further utilization, or the separated air gases can be removed from the system. Gas separation from the DCC outlet stream can be carried out in various methods. The axial flow cyclone separator is the most commonly used gas-liquid separation method widely used in industries. Kou et al. [28] simulated and experimentally proved gas-liquid separation using an axial flow cyclone separator. The experiment is conducted by passing the gas-liquid mixture stream into the cylindrical axial flow separator. A guide vane at the bottom of the cyclone separator produces centrifugal force in the fluid passing through it. Once the fluid passes into the separator, the centrifugal force created in the fluid separates the gas and liquid due to the density difference. While the liquid is collected at the bottom, the gas escapes through the top of the separator [28,29]. Ji et al. [29] experimentally proved that the efficiency of the cyclone separator could be improved by combining components to the cyclone separator. The combined cyclone separator includes components such as a steady flow element, leaf grind element, and folding plate element, which increases the efficiency of gas-liquid separation by more than 95%. Chemical looping is one of the methods of splitting the H_2O and CO_2 in the exhaust gas. The exhaust, which consists of H_2O and CO_2 undergoes a chemical reaction with the metal oxide used in chemical looping and produces different components. Farooqui et al. [30] state the process of chemical looping with cerium oxide (CeO_2). The H_2O and CO_2 are pressurized, and the temperature is raised up to 500 °C by compression. By integrating chemical looping, oxidation occurs with CeO_2, which splits H_2O into Hydrogen and CO_2 into carbon monoxide. The separated components from the exhaust of the chemical looping is further used for dimethyl ether (DME) production. This is considered one of the methods for CCU technology using chemical looping. Wotzka et al. [31] presented the possibility of separating CO_2 and water with the application of a microporous membrane. The separation of carbon dioxide and water using an MFI zeolite membrane treated with amine is analyzed experimentally and with molecular simulation. For experimental purposes, the liquid water is heated up to 120 °C, mixed with CO_2 in an evaporator, and further sent to separation. The performance of membrane separation is analyzed under different factors.

3. Experimental Facilities for Direct Contact Condensers Investigation

3.1. Description of Experiments

When considering direct contact condensers, researchers were directed to several topics. The first one covers various physical aspects of the direct contact condensation phenomenon in different construction variants, as in downcommerless trays for the steam–water system, direct contact condensation in a moving steam-water interface, in the case of the steam jet in subcooled water flow in a rectangular mix chamber or in a vertical square cross-section pipe. Additionally, visualization studies involving high-speed cameras were presented. There can also be distinguished direct contact condensation in the presence of non-condensable gas.

Other studies were devoted to analyzing heat transfer coefficient or volumetric heat transfer in direct contact condensers. Then, various construction aspects and their impact on condenser performance were analyzed. Finally, a few cases of DCC performance in the presence of non-condensable gases were given. Their short description, with emphasis on applied fluids (liquids and gases), is shown in Table 1.

In the following paragraphs, experimental studies on direct contact heat exchangers are presented. They are grouped according to the previous section into the parallel flow, counter flow, and ejector flow condensers.

Table 1. Main areas investigated in experimental studies on direct contact condensers.

Author(s)	Fluid(s)	Capacity	Operating Conditions	Remarks
Zong et al. [32]	Steam–water	Water mass flux at nozzle outlet $6\text{--}18 \times 10^{-3}\,\text{kg/m}^2\,\text{s}$	$P_{water} = 0.1\text{--}0.5$ MPa, steam mass flux at nozzle throat 200–600 kg/m^2 s	The flow field was filmed. Proposed an empirical correlation for the average heat transfer coefficient calculation
Xu et al. [33]	Steam–water	Maximum steam flow rate is 0.03 kg/s.	Steam inlet pressure 0.2–0.7 MPa Steam inlet temperature 110–170 C	Five types of plume shapes were identified visually
Mahood et al. [34]	Pentane, liquid-, vapour-water	Mass flow rate < 0.38 kg/min	Temperatures < 50 °C	Mass flow rate ratio has a significant effect on the direct contact condenser output
Ma et al. [35]	Pure steam, steam-Nitrogen and steam-Argon	Coolant mass flow rate: 0 to 8.5 t/h	Pressure in the primary loop 0.2–3.1 MPa The temperature in the primary loop 123–237 °C	The condensation heat transfer coefficient increased with pressure,

3.2. Parallel Flow Condensers

The Thermochemical Power Group (TPG) at the Polytechnic School of the University of Genoa developed and implemented the contact condenser test rig [36–39] (Table 2). It was intended for studies on the humid air turbine cycle where water introduction in a gas turbine circuit is provided by a pressurised saturator (i.e., humidification tower or saturation tower).

Table 2. The equipment in the test rig of the TPG [36–39].

Device	Rating Parameters	Comments
Water pump	0.75 kW	Centrifugal
Recirculation pump	0.33 kW	Centrifugal
Water heater	7.5 kW	Electric
Air compressor capacity	10 g/s	Maximum

Changing input variables at the levels given in Table 3, the authors performed 162 tests in total. Data analysis was provided using two types of correlations for the non-dimensional outlet air temperature, i.e., based on polynomial correlation:

$$T_{adim} = c_0 + c_1 x_1 + c_2 x_1^2 + c_3 x_1^3 + c_4 x_2 + c_5 x_2^2 + c_6 x_2^3 + \ldots, \tag{1}$$

and applying non-dimensional parameters, as non-dimensional temperature (T^*), mass flow (M^*) and the Reynolds number of inlet air (Re^*):

$$\Delta T_{adim}^* = 4.5259 \times (M^*)^{0.0326} \times (T^*)^{-0.4645} \times (Re^*)^{-0.1027}, \tag{2}$$

and:

$$\Delta T_{adim}^* = 4.8198 \times (M^*)^{0.0277} \times (T^*)^{-0.4667} \times (Re^*)^{-0.1108}. \tag{3}$$

Table 3. Experimental conditions in [36–39].

Quantity	Range	Unit
System	Air-steam	
Column diameter	80	mm
Column height	200–1800	mm
Inlet air flow	5, 7.5 and 10	g/s
Inlet water flow	5, 7.5 and 10	g/s
Air temperature	100, 200 and 300	°C
Column pressure	3, 4, 5	bar

Equation (5) was derived based on enthalpy balance and assuming an adiabatic saturation process. The standard deviation of 2.5 K and 2.8 K were obtained in the first and second case, respectively. Hence, presented relationships can be used when designing structured packing saturators.

In the work [39] the same rig was used to validate the numerical code TRANSAT developed to simulate the transient performance of direct contact heat exchangers. The error for water temperature was less than 1%.

Zare et al. [40] analysed a steam-water system with a vertical square cross-section pipe supplying equipment (Table 4). The high-speed camera (set at 100 fps) photographed the studied phenomenon.

Table 4. The test rig equipment in the study of Zare et al. [40].

Device	Rating Parameters	Comments
Water pump	1 kW	Centrifugal
Water tank	0.5 m^3	-
Water heater	6 kW	-
Steam generator capacity	90 kg/h	Maximum
Steam generator pressure	5 bar	Maximum

Based on experimental data and employing a genetic algorithm, authors developed an empirical correlation for the average heat transfer coefficient steam-water der direct contact condensation:

$$Nu_{av} = \frac{h_{av}D}{\lambda_w} = 2083 \times B^{1.47} \left(\frac{G_0}{G_m}\right)^{-1.51} \times Re^{0.525} \tag{4}$$

with:

λ_w—thermal conductivity of water, W/(m·K),
D—hydraulic diameter of the test section, m,
B—condensation driving potential, -
G_m—critical steam mass flux, G_m = 275 kg/m^2 s at an atmospheric condition
G_0—Steam mass flux, kg/m2 s.

Under experimental conditions (Table 5) the calculated average heat transfer coefficient was within the range of 0.716–3.131 MW/(m^2 K).

Table 5. The test conditions in [40].

Parameter	Value/Unit
Cross section of the test section	8 × 8 cm
Height of the test section	50 cm
Water flow rate	1–7 m^3/h
Water temperature	20–50 °C
Steam pressure	0.05–0.4 MPa
Steam temperature	108–146 °C
Steam mass flux	200–540 kg/m^2 s

Datta et al. [41] investigated direct contact condensation during subcooled water injection into a horizontal pipe supplied with steam. The test section was made of stainless steel. A steam tank was used as a steam accumulator (Table 6). Three pressure and five temperature sensors were mounted along the test section to provide their temporal variations.

Table 6. The test rig equipment in [41].

Quantity	Value	Unit
Test column diameter	66.65	mm
Test column length	2050	mm
Steam generator capacity	200	kg/h
Maximum steam pressure	16	bar
Steam tank length	1070	mm
Steam tank diameter	343	mm

During experiments, steam and water pressure varied from 2 to 3 bar and from 3 to 6 bar, respectively. Supplying water temperature was maintained about 30 °C. Steam temperature was from 120.2 °C to 133.5 °C. Authors observed higher pressure peaks (up to 6.08 bar) in the test section when the rising pressure difference between its inlet and water section was up to 3 bar.

Karapantsios et al. [42,43] considered a steam—air system using a vertical and transparent column 2660 mm high with 50 mm of internal diameter. They divided the column into the inlet (300 mm), intermediate (900 mm), and measurement (1400 mm) sections. Inlet steam pressure was maintained constant at 1.5 bar. The water flow rate was changed within the range of 26 to 416 g/s.

They defined a condensation heat transfer coefficient:

$$h_c = \frac{L}{\Delta T}\frac{\Delta W_c}{\Delta x},$$

(5)

with:

L—latent heat, J/kg,

ΔW_c—condensation rate, kg/s·m.

ΔT—temperature difference, K,

Δx—distance between measuring points, m.

Then the cumulative heat transfer coefficient was analyzed, assuming as ΔT the logarithmic mean temperature difference. During experiments, it varied between 500 and 2000 W/m^2 K for the entire condensing region (at heights between 0 and 690 mm from the steam entry).

The steam condensation subatmospheric conditions in the concurrent flow packing tower were investigated by Chen et al. [44]. As the direct contact condenser, they used a stainless steel 1000 mm high column with 300 mm of internal diameter. A steam generator with 0–144 kW of heating power and a 1.5 kW vacuum pump (6×10^{-2} Pa and flow rate of 15 L/s) were used.

During an impact of steam temperature T_{cond}, steam flow G_{in}, inlet water temperature T_{in}, and water flow L_w on the condensation process was studied. These parameters were set at values given in Table 7.

Table 7. Operating conditions during experiments in the study of Chen et al. [44].

Parameter	Value
T_{cond} (°C)	47.5, 50.0, 52.5, 55.0, 60.0
G_{in} (kg/h)	68.6, 74.6, 80.6, 86.6, 92.6
T_{in} (°C)	22–32 (0.5 interval)
L_w (m^3/h)	2.40, 2.15, 1.90, 1.65

The authors analyzed several parameters, such as condensation rate (R), degree of subcooling (ΔT), number of liquid-phase heat transfer units (NTUL), and the total volume heat transfer coefficient K_V. The relationship gives the latter one:

$$K_V = \frac{c_{pL} G_L \left(T_{out} - T_{in}\right)}{\Delta T_m \, V},$$ (6)

with:

c_{pL}—specific heat at constant pressure, J/(kgK),
G_L—mass flow rate of cooling water, kg/s,
T_{in}—inlet temperature of cooling water, K,
T_{out}—outlet temperature of cooling water, K,
ΔT_m—logarithmic average temperature difference during condensation, K,
V—volume from the liquid distributor to the stable liquid level of the tower bottom, m^3.

During experiments, K_V varied from 80 to 250 kW/(m^3 K). After fitting to experimental data, authors provided the correlation in the form:

$$K_V = 50(T_r F_{LG})^{-1.512},$$ (7)

with F_{LG} given by the equation:

$$F_{LG} = \frac{G_L}{G_{S,in}} \sqrt{\frac{\rho_S}{\rho_L}}.$$ (8)

Ma et al. [35] analyzed steam condensation in the presence of non-condensable gas (steam-nitrogen and steam argon) in relation to pure steam and estimated the heat transfer coefficient in relation to various conditions (pressure, gas content). The test section was the 1669 mm high tube with 5 mm and 8 mm inner and outer diameters, respectively. It was placed in a cylindrical container, 3660 mm high and with an internal diameter of 40 mm.

The test section was the tube-in-tube type, made of an inner tube with an outer diameter of 34 mm and an outer condensing tube with an inner diameter of 60 mm, and located in the axial centre of a stainless steel vessel with an inner diameter of 0.4 m and a height of 3.66 m. At its bottom were electrical heaters with power controlled from 0 to 60 kW, used to heat water. The condensing section had a height of 1.669 m. The thickness of the inner and outer tubes was 5 mm and 8 mm, respectively. A 60 kW electric heater was used to heat up water. The mass fraction of N_2/Ar was between 5% and 90%. Other experimental conditions are given in Table 8.

Table 8. The test conditions during experiments during pure steam condensation and steam condensation with Nitrogen/Argon.

Parameter	Steam	N/Ar	Unit
Pressure in the primary loop	0.21–3.12	0.21–4.12	MPa
Temperature in the primary loop	123–237	80–267	°C
Pressure in the coolant loop	2.01–2.46	0.40–3.2	MPa
Temperature difference between the inlet and outlet of the condensing section	9.0–12.6	10–20	°C
Average temperature difference between the primary loop and the coolant loop	45.1–80.6	27–83	°C
Pressure in the primary loop	0.21–3.12	0.21–4.12	MPa
Temperature in the primary loop	123–237	80–267	°C

During pure steam condensation, when increasing the bulk pressure from 0.21 MPa to 3.12 MPa, the average condensation heat transfer coefficient, h_c, increased from 1.74 kW/(m^2 K) to 8.95 kW/(m^2 K). The introduction of non-condensable gases significantly influenced obtained results. In the presence of N_2 with a mass fraction of 11.5% h_c increased from

0.99 kW/(m^2 K) to 4.37 kW/(m^2 K) under the system pressure rise from 0.21 MPa to 3.11 MPa. They also noticed that under constant pressure, the condensation heat transfer coefficient decreases along with the increase of non-condensable gas share. They gave a case for 4.12 MPa bulk pressure and the mass fraction of N_2 increased from 8.41% to 81.5% when h_c decreased from 6.12 kW/(m^2 K) to 0.54 kW/(m^2 K).

3.3. Counter Flow Condensers

Genic [45] analyzed a water and steam system (Table 9) comprising a 300 mm diameter column for water deaerators with downcommerless trays.

Table 9. The test conditions during experiments in the study of Genic [45].

Quantity	Range	Unit
System	Steam-water	
Column diameter	DN 300, 323.9/309.7	mm
Water flow rates at the column inlet	3.0–13.6	m^3/h
Steam flow rate	203–1070	kg/h
Inlet water temperature	20–30	°C
Water outlet temperature	39–98	°C
Steam at the inlet	102–117	°C
Working pressure in a column	100.1–101.8	kPa

The author derived experimental correlation for the number of transfer units for the liquid phase (NTU$_L$), with a correlation ratio of 0.925 and standard deviation of 15.9%, in the following form:

$$NTU_L = 0.185 \left(\frac{G_L}{G_{S,in}} \sqrt{\frac{\rho_S}{\rho_L}} \right)^{-1.48} \tag{9}$$

with:

G_L—mass flow rate of liquid, kg/s,

$G_{S,in}$—inlet mass flow rate of steam (vapour), kg/s,

ρ_L—liquid density, kg/m^3,

ρ_S—steam (vapour) density, kg/m^3.

In the next study [46] based on the same test rig, authors investigated heat transfer during direct-contact condensation on baffle trays. They presented the experimental correlation for Nusselt number based on dimensionless numbers:

$$Nu = 5.8 \times 10^{-6} \times Re^{5/3} \times Pr^{1/3} \times Fr^{-2/3} \tag{10}$$

With a correlation ratio of 0.983 and a standard deviation of 13.3% sufficient for engineering design purposes.

In the study of Chen et al. [47], the authors investigated sonic steam jet condensation in sonic flow of water. Visualization of steam plumes was performed using a high-speed camera. Then digital image processing with Matlab software was applied. The test rig was set to provide the maximum steam flow rate of 126 kg/s (Table 10). The test conditions are given in Table 11.

Table 10. The additional equipment in the test rig of Chen et al. [47].

Device	Rating Parameters	Comments
Test section length	200 mm	
Inner diameter of steam nozzle exit	5 mm	
Diameter of the restricted channel	26 mm	
Steam generator heater	90 kW	Maximum
Steam generator capacity	35 g/s	Maximum
Steam generator pressure	0.7 MPa	Maximum

Table 11. The test conditions during experiments of Chen et al. [47].

Parameter	Value/Unit
Steam inlet pressure	0.16–0.55 Mpa
Steam inlet temperature	115–155 °C
Steam mass flux	kg/(m^2 s) 200–800
Water flow rate	kg/s 0.13–0.80
Water inlet temperature	50–70 °C

The authors distinguished five considered plume shapes: hemispherical, conical, contraction-expansion-contraction, ellipsoidal, and divergent, and presented a 3-D map of these shapes.

Introducing, as in previous studies, several dimensionless parameters authors derived experimental correlation for average heat transfer coefficient:

$$h_{av} = 3.51 \times 10^{-3} \times c_p G_m B^{0.64} \left(\frac{G_e}{G_m} \right)^{-1.25} \times Re^{0.15} \qquad (11)$$

Under presented experimental conditions, h_{av} varied Nusselt number:

$$Nu_{av} = \frac{h_{av} d_e}{\lambda_w} = 0.008 B^{1.15} \left(\frac{G_e}{G_m} \right)^{-1.34} \times Re^{0.16} \qquad (12)$$

with:

d_e—diameter of steam nozzle exit, mm,

G_m—critical steam mass flux,

G_e—steam mass flux, kg/m^2 s.

The presented model produced results with an accuracy of 20% when comparing the experimental data. The measured heat transfer coefficient was within the range of 1.6–5.5 MW/m^2 K.

Fei [48] and Xu [49] investigated bubbles' uniformity and mixing time in a direct contact heat exchanger. They used a two-component system with heat transfer fluid (HTF) and R-245fa under the test conditions presented in Table 12.

Table 12. The test conditions during experiments of Fei [48] and Xu [49].

Quantity	Range	Unit
Column diameter	480	mm
Column height	1500	mm
Flow rate of HTF	0–0.3	kg/s
Refrigerant flow rate	1–3	$\times 10^4$ m^3/s

The authors concluded that there was a linear relationship between the flow patterns of a bubble swarm and heat transfer.

Xu et al. [33] investigated direct-contact condensation of the steam jet in water flow in pipes. The authors investigated the average heat transfer coefficient and Nusselt number. They also observed the plume's shape and length using a high-speed camera. The test conditions are given in Table 13.

Table 13. The test conditions during experiments of Xu et al. [33].

Quantity	Range	Unit
Height of the test section	mm	2000
Inner diameter of the test section	mm	80
Steam inlet pressure p	MPa	0.2–0.7
Steam inlet temperature Ts	°C	110–170
Steam mass flux Ge	kg/m^2 s	150–500
Water flow rate Q	kg/s	0.14–6.65
Water temperature Tw	°C	20–70

Finally, they presented experimental correlations to obtain average heat transfer coefficients. For Reynolds numbers from 2456 to 29,473:

$$h_{av} = 0.61 C_p G_m B^{0.59} \left(\frac{G_e}{G_m} \right)^{-0.58} \times Re^{0.30}. \tag{13}$$

For $29{,}473 \leq Re < 117{,}893$:

$$h_{av} = 7.21 \times 10^{-5} C_p G_m B^{0.35} \left(\frac{G_e}{G_m} \right)^{-0.55} \times Re^{1.10}. \tag{14}$$

B—condensation driving potential
C_p—water-specific heat, J/kgK
G_e—steam mass flux at nozzle exit, kg/m^2 s
G_m—critical steam mass flux, kg/m^2 s
Its value during experiments was within the range of 0.34–11.36 MW/m^2 K.

Mahood et al. [34] presented an experimental test facility for the investigation of a three-phase direct contact condenser using three phases (pentane, liquid-, vapor-water). A test section was built in the form of a 70 cm high Perspex vertical column with a 4 cm internal diameter and with seven thermocouples located along its height. The initial dispersed phase (liquid pentane) and continuous phase (water) temperature were from 37.6 °C to 41.7 °C and 19 °C, respectively.

The authors studied the impact of the mass flow rate ratio and temperature of the dispersed phase on the outlet conditions of the condenser and found that they depend mainly on the relation between dispersed and continuous mass flows. Additionally, the water temperature increased along with the column height.

In the next studies [50–58] they modified the test rig, locating thermocouples in different positions, and investigated the time-dependent volumetric heat transfer coefficient. In [50] they concluded that it decreases with time until steady-state conditions (at about 100 s in the considered case) are reached, according to the relationship:

$$U_v = \left(\frac{C_{pc}(1 - \alpha)\rho_c}{t} \right) ln \left[\frac{(T_{di} - T_{co}) + (\dot{m}_d/\dot{m}_c)h_{fg}/C_{pc}}{T_{di} - T_{co}} \right] \tag{15}$$

with:
C_{pc}—specific heat of continuous phase, J/kgK
α—holdup ratio, -
ρ_c—density of continuous phase, kg/m^3,
t—time, s,
T_{di}—dispersed phase inlet temperature, K
T_{co}—continuous phase outlet temperature, K,
h_{fg}—latent heat of condensation, J/kg.

Depending on the dispersed to continuous phases mass flow ratio, R, the initial value of U$_v$ was from about 150 kW/m^3 K at R = 6.5% to 780 kW/m^3 K at R = 43.7%. Further, in [53] they confirmed that its value was not dependent on the initial dispersed temperature.

In [54] the U_v was studied during the inception of the undesirable flooding phenomenon. Its value was from about 30 kW/m³ K to 60 kW/m³ K at T_{di} = 40°C to 60°C, respectively.

The next papers [55,56] were devoted to heat transfer by convection during direct contact condensation. Experiments showed that the unit heat transfer rate increased along with the mass flow rate ratio: from 100 kW/m³K to 200 kW/m³K at $\dot{m}_c$ = 0.05 kg/min to 200–400 kW/m³K at $\dot{m}_c$ = 0.38 kg/min.

In [51] authors developed their research determining the efficiency and capital cost of this heat exchanger. The heat transfer efficiency was given as:

$$HT_{eff} = \frac{T_{co} - T_{ci}}{T_{d,sat} - T_{ci}} \times 100\%,$$ (16)

with:

T_{ci}—continuous phase inlet temperature, K

T_{co}—continuous phase outlet temperature, K,

$T_{d,sat}$—vapour saturation temperature, K.

It was found that efficiency was controlled by means of the mass flow ratio (R). At higher values of R, the efficiency was above 50%.

This test rig with the different columns, 100 cm high with a 10 cm internal diameter, was also used [57] in investigations of the temperature distribution in the column condenser. The presented results showed a decrease in the continuous phase temperature down the height of the column. In [58,59] heat transfer measurements were performed depending on various parameters. The authors showed that the water (continuous phase) flow rate significantly affected the average volumetric heat transfer coefficient. During tests its value varied within the range of 20–60 kW/m³ K.

Observations of the transient behavior of a steam-water system with a packed column 1045 mm high and with an internal diameter of 325 mm were presented in [60]. The flow rate of cooling water was set at 120 L/h, 160 L/h, 350 L/h, 540 L/h, and 840 L/h, at a constant temperature of 28 °C.

The authors defined the volumetric heat transfer coefficient by the following equation:

$$h_v = \frac{Q_{water}}{V_e \times \Delta T_m},$$ (17)

with:

V_e—effective heat transfer volume of the column, m²,

ΔT_m—logarithmic mean temperature difference, K.

They reported that h_v increased from 1.47 kW/m³ K to 10.93 kW/m³ K with an increasing water flow rate from 120 L/h to 840 L/h. Additionally, time constant, referred to as the maximum attenuation of steam, shortened from 75 s to 13 s with the water flow rate rising within the same range.

Pommerenck et al. [61] used steam with volatile oils entrained in an air flow in the direct condenser with the sprayed water to analyze the recovery phenomenon for such oils. The condenser was built as a vertical PVC pipe with a 10 cm diameter. Water sprayer tips were mounted opposite at the same height. There were between 2 to 8 spray tips used during tests. Steam was introduced into the condenser through 2, 4, or 8 sprayers. The authors found that the direct contact condenser allowed better recovery in relation to the shell condenser. The direct condenser capture efficiency was found to be less depended on steam concentration than spray development.

3.4. Ejector Condensers

Yang et al. [32,60] investigated flow patterns and the influence of inlet water and steam parameters on pressure and temperature distributions in an ejector condenser under experimental conditions given in Table 14.

Table 14. The experimental conditions in [32,60].

Parameter	Value	Unit
Inlet steam pressure	0.1–0.5	MPa
Inlet water pressure	0.1–0.5	MPa
Steam mass flux at the nozzle throat	200–600	$kg/m^2\ s$
Water mass flux at the nozzle outlet	$6–18 \times 10^{-3}$	$kg/m^2\ s$
Inlet water temperature [32]	293–333	K
Inlet water temperature [60]	288–333	K

The authors presented selected pressure and temperature distributions for stable and unstable flow patterns, providing input conditions.

Kwidzinski [62] investigated two-phase steam-water injectors. Four devices were used, each with different dimensions. During experiments motive steam pressure was from 60 to 430 kPa with flow rates from 75 to 130 kg/h. The water flow rate was between 1500 and 6500 kg/h at a water temperature of 14 to 40 °C. The average heat transfer coefficient for condensation in a mixing chamber of the condenser was given by:

$$\alpha_{MC} = \frac{\dot{m}_{c2}(h_{V1} - h_{L2})}{A_{MC}\Delta T_{MC}},$$

(18)

with:

$\dot{m}_{c2}$—mass flow rate of condensate at the outlet of a mixing chamber, kg/s,

h_{V1}—steam enthalpy at the steam nozzle outlet, J/kg,

h_{L2}—liquid enthalpy at the mixing chamber outlet, J/kg,

ΔT_{MC}—logarithmic mean temperature difference between the vapour and liquid in the mixing chamber, K,

A_{MC}—surface area of the mixing chamber wall, m^2.

Depending on the device, it was found that α_{MC} varied from 250 kW/m^2 K to about 800 kW/m^2 K at the temperature difference (ΔT_{MC}) from 24 K to 68 K. Other experimental studies devoted to steam-ejector condensers were presented by Shah et al. [63–65]. The authors evaluated the effect of the mixing section length (110, 130, and 150 mm) on the transport process in the condenser. In addition, CFD simulations were performed.

As the source of steam, the electric 36 kW steam boiler with a 38 L tank generates saturated steam at a maximum flow rate of 52 kg/h (14.4 g/s) and pressure of 8 bar. The operating conditions during experiments are given in Table 15.

Table 15. The experimental parameters of steam, water, and pressure in [63–65].

Parameter	Value	Unit
Steam inlet pressure	140–220	kPa
Steam inlet temperature	382–396	K
Water inlet pressure	96	kPa
Water inlet temperature	290	K
Ambient pressure	96	kPa

The authors didn't present experimental correlations. However, several general outcomes were given. They observed an increasing water mass flow rate along with increasing inlet steam pressure. Additionally, under the same operating conditions, they obtained higher suction pressure and flow rate at a shorter length of the mixing section.

Reddick et al. [66] investigated a steam ejector's performance in a mixture of steam and carbon dioxide (as non-condensable gas). The test rig included the 75 kW electric boiler (maximum pressure of 600 kPa), a 3 kW superheater, an ejector, a flash tank, and a condenser. Several operation variants were considered (Table 16) and performance curves were then prepared.

Table 16. The experimental conditions in [66].

Parameter	Value	Unit
Primary inlet pressure	350, 450, 550	kPa
Secondary in let pressure	50, 70, 90	kPa
Nozzle throat diameter	4.03, 4.23, 4.59, 5.09	mm

When pure steam, without CO_2 entraining, was used, then it was observed that increasing the primary pressure, secondary pressure or nozzle diameter resulted in a higher value of critical pressure and lower critical entrainment ratio.

Entraining CO_2 resulted in different outcomes. Authors reported that the rising share of CO_2 resulted in an increased critical entrainment ratio, linearly. At the same time, the critical pressure was unchanged.

4. Measuring Systems in DCC Analysis

The main physical quantities measured in test rigs with direct contact condensers include temperature, pressure, and flow rate. A large variety of measurement methods and techniques can be applied here. However, from a practical and economical point of view, those that the authors found best in a given case were used. It should be emphasized here that none of the presented publications gave reasons for this or that choice. So, it may be worth giving a short presentation of various measurement methods with their main advantages and disadvantages and then presenting a short summary of findings.

The first criterion used when choosing a given sensor is based on the design of the test rig. From this, one can say if there is an electronic data acquisition system or not. If so, a sensor with an electric output signal should be used. If not, there can be applied simpler and cheaper solutions. Scientific experiments require data measurement and acquisition for further processing. Therefore, the presented description covers mainly measurement sensors with electric output signals, which can be used in modern data acquisition systems.

Regarding the temperature measurement, the temperature range of process fluids in the presented papers has not exceeded 300 °C [36]. Hence, thermocouples (TC) and resistance temperature detectors (RTD) could be useful. This is so because these kinds of sensors provide electrical-type output measurement signals that can be very easily transmitted and converted into computer measurement systems. Thermocouples can be used for a wide measurement range, from –270 °C to +1370 °C (K-type chrome–alumel thermocouple) to over 2000 °C (Pt-Rh thermocouples). The IEC 60584 standard defines classes 1 and 2 of tolerance. For class 1 of a measured temperature of 200 °C tolerance is from ±0.50 °C (type T) to ±1.50 °C (types E, J, K, and N). Due to the fact that the sensitive measuring point of the thermocouple (measuring junction) can be very small, with a diameter below 0.5 mm, the response time of these sensors can be very short. Protection of this junction against the negative impact of the external environment in a protective tube (sheath) results in a greater value of this response time [67,68].

The second kind of electrical temperature sensor are RTDs, which are more expensive, larger, and more fragile than thermocouples. Yet they have good accuracy, stability, and sensitivity [67,68]. At a temperature of 200 °C, according to the IEC 60751, the wire-wound A-class Pt100 sensor has a tolerance of ±0.44 °C. Protection of the sensitive part, platinum resistive wires, against the negative environmental impact results in a longer thermal time constant. Temperature measurement in direct contact condensers in presented test rigs was performed mainly by thermocouples. Additionally, RTDs were used but on a smaller scale. Their main parameters are given in Table 17.

Table 17. Sensors for temperature measurement are used in experimental rigs with direct-contact condensers.

Type	Diameter [mm]	Accuracy	Reference
Pt100	n.a.	±0.1 °C	[47]
Pt100	n.a.	0.1 °C	[46]
Pt100	n.a.	0.1 °C	[69]
K	n.a.	±1 K	[50–56]
K	n.a.	±1.5 K	[39]
T	n.a.	0.1 °C	[42]
K	1.0	0.5 K	[62]
Pt100	0.48 mm	±0.1 °C	[66]
K	n.a.	1 °C	[64]
J	n.a.	0.5%	[41]
K	3.0	0.75%	[35]

As pressure measurement is considered, dominated piezoresistive absolute and differential pressure transducers [48,50], and piezoelectric transducers were used for pressure measurements. [49,66]. They have good accuracy and sensitivity.

The next very important measured quantity is the flow rate of various liquids and gases in the presented test rigs. Genić et al. [37] applied orifice flow meters manufactured following ISO 5167-1, with classical mercury U-tube manometers. When the water flow rate is to be considered, the most popular solution was the electromagnetic flow meter. However, in several cases, the turbine flow meter was also chosen (Table 18).

Table 18. Devices used in water flow measurements.

Type	Range	Error	Reference
Rotameter	1–7 m^3/h	0.1 m^3/h	[40]
Electromagnetic	0.08–2.78 kg/s	0.2%	[60]
Electromagnetic	0.88–17.66 m^3/h	0.5%	[47]
Electromagnetic	0–10 m^3/h	1.0%	[41]
Rotameter	-	1.25%	[56]
Turbine	0.04–0.25 m^3/h	1.0%	[47]
Turbine	0.6–6 m^3/h	1.0%	[47]
Rotameter	1–10 m^3/h	1.5%	[47]
Turbine	0.9–13.6 m^3/h	0.15%	[47]

In steam flow measurement, vortex flow meters were the most popular (Table 19). However, despite the wide range of analysed studies, no selection guidelines were given by the authors.

Table 19. Devices used in steam flow measurements.

Type	Range	Error	Reference
Orifice	-	1.0 Pa	[64]
Vortex	7.5–73.9 g/s	1.0%	[57]
Vortex	30 to 300 m^3/h	1.5%	[59]
Vortex	0–40 m^3/h	0.75%	[28]
Vortex	0–120 m^3/h	0.75%	[28]
Vortex	0–150 kg/h	1.0%	[59]
Vortex	-	2.0%	[61]
Orifice	-	1.5%	[61]

The presented review shows various measurement techniques used in experiments with direct contact condensers. Despite the importance of this issue, authors presented a general overview and review about the selection of measurement equipment devoted to the test rig with the ejector condenser.

5. Numerical Modeling of Direct Contact Condensation

Numerical modeling of Direct Contact Condensation (DCC), which occurs in Direct Contact Condensers (DCCs), is challenging because of the phenomenon's complexity. It requires taking into account the multiphase flow, often combined with turbulence. Computational Fluid Dynamics (CFD) can fully overcome these challenges, which is the most common tool for DCC modeling. There is no universal modeling framework for DCC because of the diversity of the phenomenon. Various flow regimes can occur (stratified flow, bubbly/droplets flow, etc.). The Euler-Euler interface tracking methods are suitable for liquid/vapour jet-type condensation, where the heat and mass transfer occurs mainly on the interface between phases. The most known clear interface tracking method is the Volume of Fluid (VOF). For drop-type direct condensation, where the vapor condenses on the surface of the droplets, methods allowing for a dispersed phase should be used. They can be based on the Euler-Euler and Euler-Lagrange approaches [69], but the Euler-Euler approach is often used. In this framework, two models are worth mentioning: the Mixture and the Eulerian two-fluid model. The Eulerian two-fluid model allows for modelling a mixed flow regime combined with interface tracking. Together with the k-ε model, it is often used for modeling steam/bubble jets submerged in subcooled water [70]. The major disadvantage of this model is extensive computational time. This section contains examples of numerical modeling of Direct Contact Condensation in various types of devices.

5.1. Numerical Analysis of DCC of Vapour Injected in the Liquid Tank

The direct contact condensation of steam from a vertical pipe into a water pool was examined numerically by Kunwoo Yi et al. [71]. The scheme of the geometrical model was presented in Figure 6. The presence of inert gas (air) was taken into account. The Star CCM+ solver was used. The Volume of Fluid model (VOF) was used to simulate multiphase flow. Reynolds-Averaged Navier-Stokes equations were closed using the k-ω SST model. The mixture of gas (steam and air) was assumed as an ideal gas. The evaporation/condensation model was used to model the direct contact condensation phenomenon. Mesh consists of mesh with 3 million polyhedral elements. The steam rate was 0.468 kg/m² s, and the flow was chugging. The tank was initially filled with water in 30% above the bottom of the suppression pool. The initial pressure in the suppression pool was atmospheric, and the temperature was 48.9 °C. As a result of the study, the influence of vertical tube pin holes on the behavior of chugging flow was investigated. The low steam flow rate in the blowdown pipe can prevent the chugging flow.

Multiphase CFD analysis of steam-water direct contact condensation in a Pressure Suppression Chamber was conducted by Tyler Dee Hughes [73]. The pressure Suppression Chamber is a crucial part of the BWR Reactor Core Cooling system. A 2D, axisymmetric model was developed using commercial Star CCM+ software based on the Finite Volume Method (FVM). The Eulerian two-fluid multiphase model with a segregated solver was used. Direct contact condensation was modelled based on the Hughes-Duffey Nusselt number correlation correlated to the liquid side:

$$Nu = \frac{2}{\pi} Re_t^{1/2} Pr_l \qquad (19)$$

with:

Re_t—turbulent Reynolds number, -,
Pr_l—liquid Prandtl number, -,

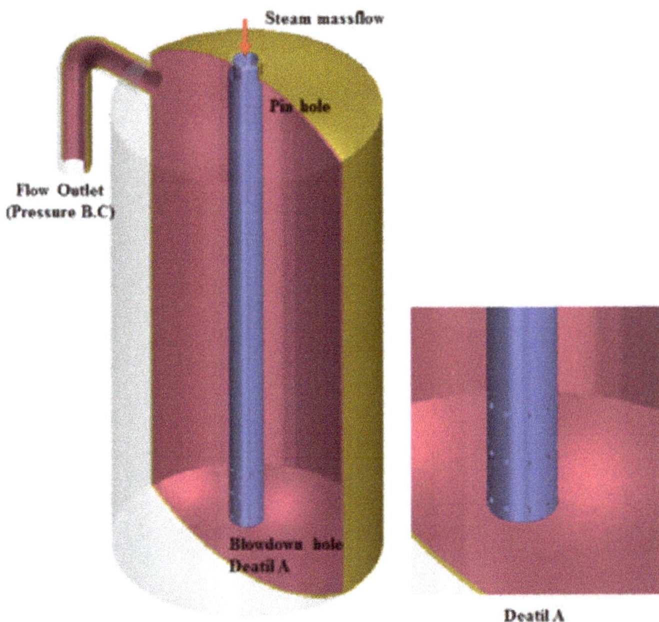

Figure 6. Suppression pool geometrical model [72].

The Standard k-ε turbulence model was used to model both phases' behavior. The boundary conditions were assumed in accordance with the experimental environments. The physical properties of water and steam were computed based on IAPWS tables [74]. The simulation was transient First-order backward Euler implicit time step discretization and was applied; the Courrant Number did not exceed 10. For all fields, the second order discretization scheme was used. Simulation relaxation factors were presented in Table 20. Two types of meshes are considered: polygonal and structured. The temperature of the steam was 120 °C (Saturation temperature for 197 kPa pressure), and the mass flow was 34 kg/m^2 s. The average water temperature in the pool was 67 °C. The bubbling flow regime was observed numerically and experimentally. Figure 7 shows the condensation which occurs near the periphery of the bubble. In the simulation, rapid changes in the pressure were observed. The 2D axisymmetric structured mesh was the best for this type of calculation.

Table 20. Simulation relaxation factor [71].

Solver Field	Relaxation Factor
Phase coupled velocity	0.56
Pressure	0.2
Volume Fraction	0.1
Energy	0.3
Turbulence	0.3

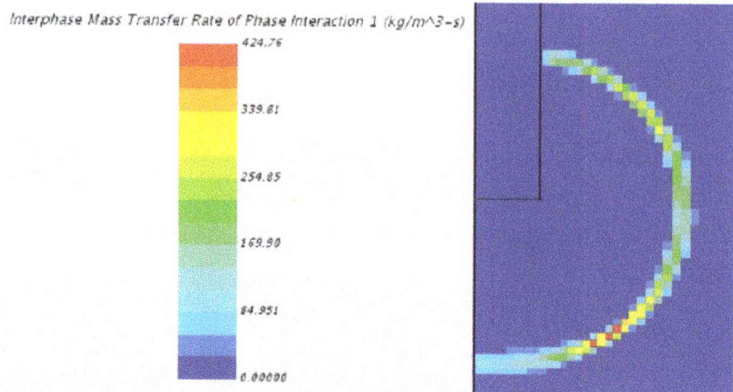

Figure 7. Condensation rate contour in case of steam injection [73].

Roman Thiele [75] did the modeling of Direct Contact Condensation of saturated steam to subcooled water using the opensource, OpenFoam framework. The condensation occurs in the suppression pool, which is part of the Nuclear Power Plant. The solver uses the VOF method based on the cavitation model earlier developed by Kunz [76]. The governing equations are based on the volume continuity law. Pressure, velocity, and phase continuity equations were solved separately. Two-phase change models were developed to describe direct contact condensation: one based on the combustion approach and the second using inter-facial heat transfer. The simulation model of the facility was presented in Figure 8. The fluid properties were based on the IF97 database [74]. The steam temperature was 102 °C, and the corresponding saturation pressure was 1.1 bar. The liquid temperature was 22 °C with the same ambient pressure. Various steam mass flow rates were used for the time step and velocity dependence tests: 1.2 g/s, 4.9 g/s, and 12.4 g/s. The interface model better predicts the direct contact condensation phenomenon. It also shows great time-step stability. It should be considered for further development.

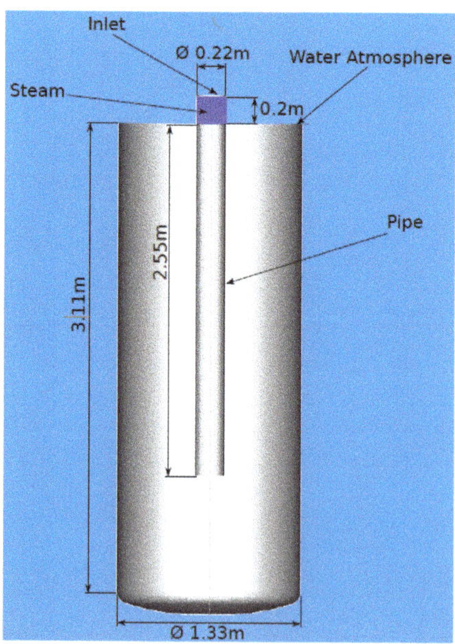

Figure 8. The model used for simulation [75].

Jayachandran et al. [77] investigated numerically bubbling direct contact condensation of gas and liquid oxygen to capture heat and mass transfer effects. Direct contact condensation takes place as a result of mixing between hot gas mixture jet and subcooled liquid oxygen in booster turbopump exit of oxidizer-rich staged combustion cycle. To simplify the complex phenomenon, the gas mixture consists only of pure oxygen. The problem was solved using CFD methods with the ANSYS CFX solver based on the Finite Volume Method (FVM). For multiphase flow modeling, the two-fluid (particle-based) model was applied. The mean bubble diameter was taken from Anglart et al. [78]. RANS approach with separate equations for phases was used in the case of turbulence modeling. The two-equations$k$-ε model was used for modeling turbulence on the liquid side, and the dispersed phase zero equation model was used for the vapor side [79]. The first-order upwind scheme was used to solve turbulence and advection schemes and the first-order backward Euler scheme to solve unsteady terms. The time step was 10^{-4} s. Direct Contact Condensation was modeled using the thermal phase change model, which is a two-resistance model. The Nusselt number for the vapor side was calculated using the zero-equation formula, and the liquid side was based on the Ranz Marshall model [80], which is expressed below:

$$Nu = 2 + 0.6Re^{0.6}Pr^{0.3} \tag{20}$$

The mass flow rate of steam and temperature were respectively 0.0051 kg/s and $-293\ ^\circ$C. The liquid temperature was 208 $^\circ$C, and the pressure in the tank was 0.1 MPa. Figure 9 presents the heat transfer coefficient for a typical cycle. The observed DCC heat transfer coefficient is approximately ten times higher than in film condensation (for oxygen). The maximum value of the heat transfer coefficient and the strongest pressure oscillations are achieved for the necking stage of bubbling DCC.

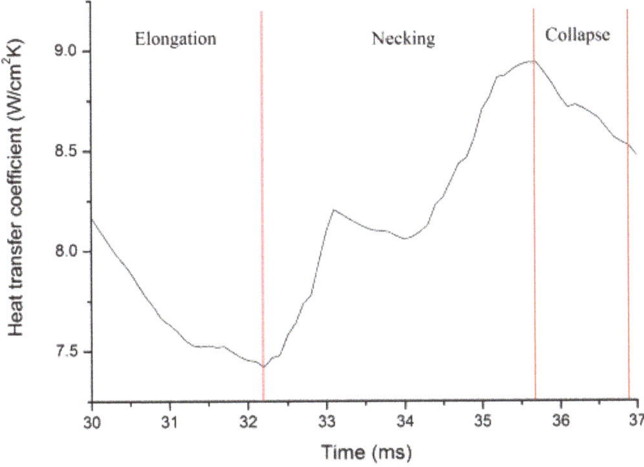

Figure 9. Heat transfer coefficient for a typical cycle [77].

5.2. Numerical Analysis of Jet-Type Flow Direct Contact Condensers

The condensing ejector was the object of the numerical investigation conducted by Colarossi et al. [72]. The aim was to develop a CFD model of condensing ejector producing growth in static pressure in refrigeration systems based on a CO_2 working medium. Computation Fluid Dynamics supplemented with semiempirical correlations was used to create a numerical model. The open-source OpenFoam library was used and the Eulerian pseudo-fluid approach was applied. The thermodynamic non-equilibrium state of the working fluid was taken into account. Pressure and velocity fields were calculated implicitly using the PISO algorithm (Pressure Implicit with Splitting of Operators). Modified versions of the

Homogeneous relaxation model (HRM) were used (originally HRM model has developed for modeling of Flash boiling phenomenon [81]). The HRM model is based on the total derivative which describes the mass fraction of vapour:

$$\frac{Dx}{Dt} = \frac{\overline{x} - x}{\theta} \tag{21}$$

with:

    x—quality (mass fraction of vapour), -,
    $\overline{x}$—equiblirum quality, -,
    t- time, s,
    θ- timescale, s,
    In considering the modified HRM model, the timescale is expressed as:

$$\theta = \theta_o \alpha^a \psi^b (1 - \alpha)^a \tag{22}$$

with:

    θ_o, a, b—model constants, -,
    α—vapour volume fraction, -,
    ψ—dimensionless pressure difference, -.

The HRM model coefficient values for low-pressure conditions (below 10 bar) were taken from Downar-Zapolski et al. [82]. RANS equations were supplemented with the k-ε model. The 2D mesh consists of nearly 6100–6750 elements was used. The results show good agreement with the experimental data. Turbulence modeling was marked as the most challenging in the case of condensing ejectors modeling.

The three-dimensional CFD model of condensing ejectors was developed by Bergander et al. [83]. The Condensing Ejector is part of the second compression step in the refrigeration cycle. The article contains theoretical, numerical, and experimental analyses. The scheme of condensing the ejector and pressure distribution is presented in Figure 10. The working medium was R22. The study aims to calculate the exit pressure of the condensing ejector at given inlet boundary conditions. Operating parameters are presented in Table 21. The temperature was taken from R22 tables based on pressure and enthalpy values presented by the authors.The pseudo-fluid approach was applied, considering a mixture of no thermodynamic-equilibrium conditions with a homogeneous relaxation model HRM. The prepared model can be used for flash-boiling and condensation modeling.

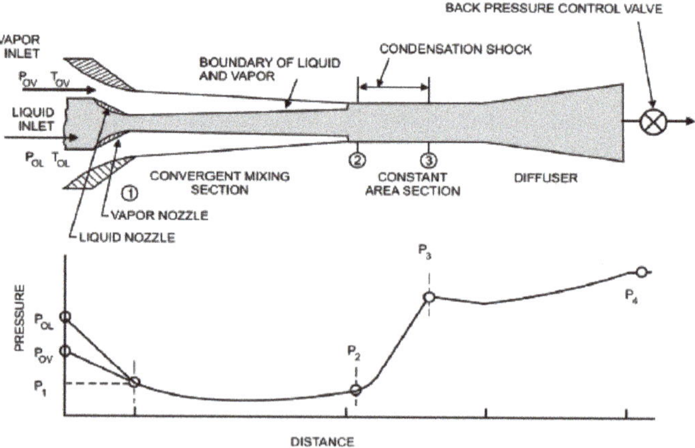

Figure 10. Condensing ejector scheme and pressure distribution [83].

Table 21. Operating parameters from Bergander et al. [83].

Parameter	Vapor Inlet	Liquid Inlet
Temperature (°C)	43	34
Pressure(MPa)	1.7	2.1
Massflow (kg/s)	0.016	0.077

Zhang et al. [84] conducted a numerical and experimental study of steam-water ejectors in a trigeneration system for hydrogen production. Direct Contact Condensation takes place in the presence of non-condensable gas (air). The aim was to predict the performance of the ejector using CFD method. CFD solver available in ANSYS 18.2 version was used. For multiphase modeling, Euler-Euler two-fluid model was used. The species transport model was incorporated into the simulation to consider the inert gas's presence. K-ω SST model with enhanced wall treatment accounted for the turbulent flow phenomenon. All equations were descretized using a high-order scheme. The thermal equilibrium model, a two-resistance model, was applied to model direct contact condensation [85]. The condensation rate per unit volume $\dot{m}$ from gas phase α to liquid phase β:

$$\dot{m} = \frac{h_\alpha A_{\alpha\beta}(T_s - T_\alpha)}{H_{S\beta} - H_w} \tag{23}$$

with:

h_α—heat transfer coefficient at liquid side, W/(m^2 K),

$A_{\alpha\beta}$—interface area per unit volume, 1/m,

T_s—saturation temperature, K,

T_α—liquid phase temperature, K,

$H_{S\beta}$—specific enthalpy of steam, J/kg,

H_w—specific enthalpy of liquid-phase at the gas-liquid interface temperature, J/kg,

The heat transfer coefficient h_α for the liquid side was calculated based on the Nusselt Number expressed by the Hughmark model [86]:

$$Nu = 2 + 0.6Re^{0.5}Pr^{0.33}, \ 0 \leq Re < 776.06, \ 0 \leq Pr < 250 \tag{24}$$

$$Nu = 2 + 0.27Re^{0.62}Pr^{0.33}, \ 776.06 \leq Re, \ 0 \leq Pr < 250 \tag{25}$$

The 3D mesh consisting of 141,376 hexahedral elements was created using ICEM CFD. The steam mass flowrate was 1.45 g/s. The temperature of the water was 8 °C. At the outlet, a 1 atm pressure boundary condition was assumed. Simulation parameters are summarized in Table 22. Properties of water and steam were assumed based on the IAPWS-IF-1997 [74]. The density of air was calculated using the ideal-gas law. For a small amount of air, the performance of the ejector was improved. The achieved maximum value of condensation rate is 3252–2340 kg/m^3 s, depending on the air concentration.

Table 22. Simulation parameters from Zhang et al. [84].

Parameter	Gas Inlet	Liquid Inlet	Outlet
Temperature (°C)	104.8 (steam) 20 (air)	9	-
Pressure(kPa)	120	100	100
Massflow (g/s)	1.45 (steam) 0–0.14 g/s (inert gas)	34.7–37.3	

Shah et al. did a numerical and experimental investigation of steam pumps taking into account the direct contact phenomenon [63]. The task of the steam jet pump is to pump water. CFD, a three-dimensional model, was developed. Ansys Fluent 6.3 software was used. The multiphase flow was modeled using the Eulerian two-fluid model. The direct-contact condensation phenomenon was considered using the two-resistance model, developed earlier by [85]. The heat transfer coefficient of the liquid side was calculated based on the Hughmark Nusselt number correlation [86]. The heat transfer coefficient for steam bubbles was calculated according to the Brucker and Sparrow formula [87]. Steam is modeled using ideal gas relation. A realizable k-e model was used. Continuity and volume fraction equations were discretizated using second-order and first-order upwind schemes, respectively, and the remaining equations using Power law scheme. Mesh consists of 69,677 hexahedral and tetrahedral elements. Various steam inlet pressures were considered: 140, 160, 180, 200, 220 kPa. The steam temperature was saturated at mentioned pressures. The different water nozzle pressures were assumed: 93.56, 92.92, 91.87, 90.38, and 89.30 kPa. The water nozzle temperature was 17 °C. The static pressure and temperature charts along the length were prepared. ThThe temperature agreement between simulation results and experimental data can be observed. Considering conditions, it is possible to suck in the water from 2.12 m depth.

A numerical investigation of condensing water jet eductorwas done by Koirala et al. [88]. Mass and heat transfer in direct contact condensation occurs in two-phase flow were computationally studied using the CFD method. The aim of the study was to investigate the performance of the device as a direct contact condenser for various operational conditions. The eductor is part of a thermal desalination system. The motive fluid was water, and the sucked-in fluid was steam. The Eulerian model using ANSYS Fluent software was used to calculate multiphase flow. Fluid turbulence was taken into account using k-ωmixture model. Inlet and outlet pressure boundary conditions were applied and presented in Table 23. For direct contact condensation heat transfer calculations, two resistance models were applied. For mass transfer calculations, the thermal phase change model was used. Pressure based double precision solver was used. Under-relaxation factors of pressure, turbulent kinetic energy, and turbulent dissipation rate were respectively 0.1, 0.2 and 0.4. As a result of the study, the influence of back pressureand motive fluid temperature on the performance of eductor was investigated. Increasing back pressure decreased the flow of entertainment. Increasing the motive fluid temperature causes a lower entertainment ratio because a decrease in the condensation rate can be observed.

Table 23. Assumed boundary conditions from Koirala et al. [88].

Parameter	Gas Inlet	Liquid Inlet	Outlet
Temperature (°C)	100	25	-
Pressure(kPa)	45, 60, 80, 105	1000	100

5.3. Other Works

CFD simulation of direct contact condenser in the presence of inert gas for Oxy-fuel CO_2 Capture process was developed by Takami et al. [89]. The scheme of the device is presented in Figure 11. The aim of the condenser is to separate steam and CO_2 through the condensation process. The goal of the investigation was to provide a better understanding of the separation process depending on various boundary conditions and different fluid properties. The process was conducted based on 2D CFD modeling using a COMSOL solver. The simulation parameters at inlets and outlets are presented in Table 24. Constant fluid properties were assumed. The triangular mesh with 27,054 elements was applied. The application of considering condenser allows for condensing 75% of water content from exhaust gases. Laminar flow regimes were observed in the middle of the condenser and zones near the walls.

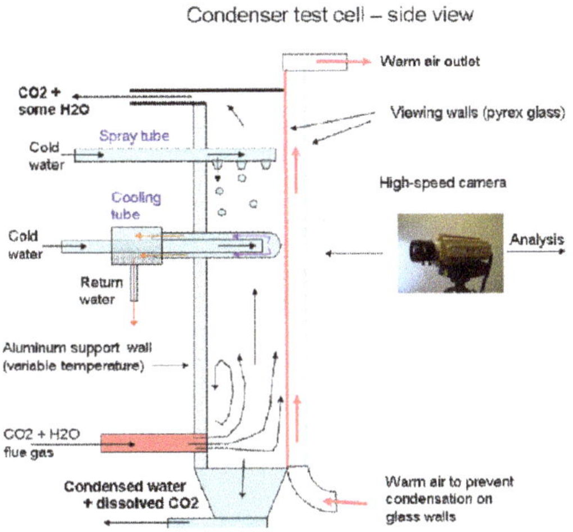

Figure 11. Direct contact condenser—scheme [89].

Table 24. Simulation results and assumptions [89].

Parameter	Gas Inlet	Gas Outlet	Liquid Inlet	Liquid Outlet
Composition (%)	$CO_2 = 92.30$ $H_2O = 5.76$ $O_2 = 1.94$	$CO_2 = 97.88$ $H_2O = 1.2$ $O_2 = 0.92$	$H_2O = 100$	$H_2O = 99.92$ $CO_2 = 0.02$
Temperature (°C)	50	27.8	25	51
Pressure(Pa)	100,000	101,325	150,000	101,325
Massflow (kg/s)	8.3	7.8	9.3	9.78

Numerical modeling of a Direct Contact Condensation of steam in a horizontal pipe was conducted by Thomas Hofne et al. [90]. The study aimed to model the two-phase stratified steam-water flow experiment and compute new heat and mass transport models between water and steam. In this stratified flow pattern, condensation occurs mainly on the interface. The Eulerian two-fluid model was used. Algebraic Interfacial Area Density (AIAD) model was implemented in the Ansys CFX solver. The models allow for simulating momentum exchange depending on the character of the stratified flow. Three various DCC models, which express the correlations for Nusselt number calculation, were used: the Egorov model with the Ranz Marshall correlation, modified Hughes-Duffey with the Ranz Marshall correlation, and the Adapted Coste model. The Nusselt Number for the last one can be expressed as the following:

$$Nu = 2.7 Re_t^{0.875} Pr_l^{1/2} \qquad (26)$$

The gas temperature was 100 °C (saturation temperature), and the mass flow rate was 5.3 g/s. The water temperature was 20 °C (initially), and the mass flow rate was 13.8 g/s. No slip boundary conditions were applied, and the simulation was stationary. Mesh consisted of 1.3 million elements.

5.4. Numerical Analysis of DCC—Summary

Tables 25 and 26 summarize the most important issues connected with the numerical modeling of direct contact condensation. Table 25 is an overview of the computational models (multiphase, turbulence, and condensation). Table 26 shows the simulation conditions in the listed research: mass flow rate, temperature, and pressure of phases. In the last column of Table 26, the main conclusions are presented. Still, the developed direct condensation CFD models need to be carefully calibrated and physically validated by the use of experimental results [91,92].

Table 25. Overview of direct contact condensation models.

Authors	Device	Phases	Solution Metod	Multiphase Model	Turbulence Model	DCC Model
Yi et al. [71]	Suppression pool	Steam-water (air)	FVM	VOF	k-ω SST model	Evaporation/ condensation
Hughes [73]	Suppression pool	Steam-water	FVM	Eulerian two-fluid	Standard k-ε	Interface method
Jayachandran et al. [77]	Suppression pool	Vapour-liquid (O2)	FVM	two-fluid (particle-based)	k-ε (liquid) 0 eq. (vapour [79])	Thermal phase change model (two-resistance model)
Thiele [75]	Suppression pool	Steam-water	FVM	VOF		combustion method, interface method
Colarossi et al. [72]	Ejector	Liquid-vapor (CO2)		Pseudo-fluid	Standard k-ε	modified HRM [82]
Bergander et al. [83]	Ejector	Liquid-vapor (R22)		Pseudo-fluid		modified HRM [82]
Zhang et al. [84]	Ejector	Steam(air)-water	FVM	Eulerian two-fluid	k-ω SST model	two-resistance model [85]
Shah et al. [63]	Ejector	Steam-water	FVM	Eulerian two-fluid	Realizable k-ε	two-resistance model [85]
Koirala [88]	Eductor	Water-steam	FVM	Eulerian two-fluid	k-ω	Thermal phase change model (two-resistance model)
Takami et al. [89]	Direct Contact Condenser	Water-Steam (CO$_2$)				
Hohne et al. [90]	Pipes	Water-steam	FVM	Eulerian two-fluid		Egorov model, Hughes-Duffey, Adapted Coste

Table 26. Overview of direct contact condensation simulation conditions.

Authors	Phases	Phase 1			Phase 2			Main Conclusion
		m [g/s]	P [kPa]	T [°C]	m [g/s]	P [kPa]	T °C]	
Yi et al. [71]	Steam-Water(air)	-	101.3	48.9	-	-	-	It is possible to prevent the chugging flow
Hughes [73]	Steam-water	45.0	197.0	120.0	-	-	67.0	2D axisymmetric structured mesh was found to be the best for this type of calculation
Jayachandran et al. [77]	Vapour-liquid (O2)	5.1	-	−293.0	-	100	−208.0	Heat transfer coefficient is approximately 10 times higher than in film condensation
Thiele [75]	Steam-water	1.2, 4.9; 12.4	110.0	102.0	-	110.0	22.0	Interface model better predict DCC phenomenon
Colarossi et al. [72]	Liquid-vapor (CO$_2$)	-	-	-	-	-	-	Turbulence modeling is the most challenging task in case of ejector modeling
Bergander et al. [83]	Liquid-vapor (R22)	77.0	2100.0	34.0	16.0	1700.0	43.0	Prepared model can be used for flash-boiling and condensation modeling
Zhang et al. [84]	Steam(air)-water	1.45, 0–0.14 (air)	120.0	104.8, 20 (air)	34.7–37.3	100	9.0	For the small amount of air, the performance of the ejector is improved
Shah et al. [63]	Steam-water	-	140, 160, 180, 200, 220	Satura-ted	100–700	-	17.0	Considering ejector allow to suck in the water from 2.12 m depth
Koirala [88]	Water-steam	-	1000.0	25.0	45, 60, 80, 105		100.00	Increasing a motive fluid temperature causes decreasing in the condensation
Takami et al. [89]	Water-Steam (CO$_2$)	9.3	150.0	25.0	8.3	100	50.0	75% of water content from exhaust gas was condensed
Hohne et al. [90]	Water-steam	13.8		20	5.3		100.0	All considering models are in good agreement with experimental data

6. Conclusions

A wide range of applications of Direct Contact Condensers indicates the strong need to investigate these phenomena occurring during the direct contact condensation process. Computational and experimental studies are two more popular ways to explore and investigate heat and mass transfer processes. To date, the developed direct condensation models still need to be carefully calibrated and physically validated by the use of experimental results. On the other side, the experimental test rig's conceptual design should be first analyzed by using available numerical results. The paper presented a comprehensive review of experimental and numerical investigations on the DirectContact Condensation process.

CFD computational methods are very helpful in the numerical analysis of direct contact condensers because of the phenomenon's complexity. Multiphase, turbulent flow with phase change requires sophisticated methods to consider all crucial aspects. Commercial software (Ansys, CFX, STAR CCM+) are most often used. Numerical calculations mainly concern cases in nuclear reactor safety systems (suppression pools) and refrigeration and heating systems (condensing ejectors). Various boundary conditions and geometries cause the flow of structure occurring during direct contact to be very diverse (jet, bubbly/droplets flows). Because of the diversity of the flow patterns, there is no universal modeling framework. Pseudofluid multiphase approaches are often used for computational calculations of ejector condensers. The VOF model is used for modeling DCC in suppression pools, where vapor is injected into stationary liquid. The eulerian two-fluid model is suitable for a wide range of applications (ejectors, tanks, pipes, etc.). Two equations RANS models: the k-ε model and the k-ω SST modelare sufficiently accurate for turbulence modelling in case of direct contact condensation. For DCC heat and mass transfer calculation, interface methods are dominating. In this model, heat transport is most often calculated based on Nusselt number correlation, which allows for calculating the heat transfer coefficient. Another type of approach, mainly used in the numerical calculation of condensing ejectors, is HRM model. It is strongly based on empirical correlations but tuned with coefficients and is characterized by robustness and sufficient accuracy. The review of the numerical investigation shows that various types of direct contact condensation modelling approaches are still developing because of the immense diversity and complexity of the phenomenon.

Various measurement methods and techniques applied to direct contact condensation experiments are presented in the studies. In temperature measurement, thermocouples prevailed due to their short thermal time constant and small dimensions. Pressure measuring techniques are based mainly on piezoelectric or piezoresistive transducers. Water flow rate measurement was performed using mainly electromagnetic and turbine flow meters. For steam flow rate measurements, mainly vortex flow meters were used. The developed detailed guidelines for measurement equipment in DCC experimental campaign are complicated and should be designed individually for the selected type of Direct Contact Heat Exchanger and operating conditions.

Author Contributions: Conceptualization, P.M. (Paweł Madejski); methodology, P.M. (Paweł Madejski), P.M. (Piotr Michalak), T.K.; formal analysis, P.M. (Paweł Madejski), P.M. (Piotr Michalak), T.K., investigation, P.M. (Paweł Madejski), T.K., P.M. (Piotr Michalak), N.S., M.K.; resources, P.M. (Piotr Michalak), T.K., P.M. (Paweł Madejski), M.K., N.S.; data curation, P.M. (Paweł Madejski), T.K., P.M. (Piotr Michalak); writing—original draft preparation, T.K., P.M. (Piotr Michalak), P.M. (Paweł Madejski), N.S.; writing—review and editing, P.M. (Paweł Madejski); T.K., P.M. (Piotr Michalak); visualization, P.M. (Paweł Madejski), P.M. (Piotr Michalak), T.K., N.S.; supervision, P.M. (Paweł Madejski); project administration, P.M. (Paweł Madejski); funding acquisition, P.M. (Paweł Madejski). All authors have read and agreed to the published version of the manuscript.

Funding: Research project supported by the program, "Excellence initiative—research university" for the AGH University of Science and Technology.

Data Availability Statement: Not applicable.

Conflicts of Interest: The authors declare no conflict of interest.

References

1. Jacobs, H.R. *Direct-Contact Condensation. InDirect-Contact Heat Transfer*; Kreith, F., Boehm, R.F., Eds.; Springer: Berlin/Heidelberg, Germany, 1988; pp. 223–236. [CrossRef]
2. Chantasiriwan, S. Effects of cooling water flow rate and temperature on the performance of a multiple-effect evaporator. *Chem. Eng.Commun.* **2015**, *202*, 622–628. [CrossRef]
3. Zhao, X.; Fu, L.; Sun, T.; Wang, J.Y.; Wang, X.Y. The recovery of waste heat of flue gas from gas boilers. *Sci. Technol. Built Environ.* **2017**, *23*, 490–499. [CrossRef]
4. Prananto, L.A.; Juangsa, F.B.; Iqbal, R.M.; Aziz, M.; Soelaiman, T.A.F. Dry steam cycle application for excess steam utilization: Kamojang geothermal power plant case study. *Renew. Energy* **2018**, *117*, 157–165. [CrossRef]
5. Sanopoulos, D.; Karabelas, A. H_2 Abatement in Geothermal Plants: Evaluation of Process Alternatives. *Energy Sources* **1997**, *19*, 63–77. [CrossRef]
6. Barabash, P.; Solomakha, A.; Sereda, V. Experimental investigation of heat and mass transfer characteristics in direct contact exchanger. *Int. J. Heat Mass Transf.* **2020**, *162*, 120359. [CrossRef]
7. Ebrahimi, M.; Keshavarz, A.; Jamali, A. Energy and exergy analyses of a micro-steam CCHP cycle for a residential building. *Energy Build.* **2012**, *45*, 202–210. [CrossRef]
8. Aidoun, Z.; Ameur, K.; Falsafioon, M.; Badache, M. Current Advances in Ejector Modeling, Experimentation and Applications for Refrigeration and Heat Pumps. Part 1: Single-Phase Ejectors. *Inventions* **2019**, *4*, 15. [CrossRef]
9. Aidoun, Z.; Ameur, K.; Falsafioon, M.; Badache, M. Current Advances in Ejector Modeling, Experimentation and Applications for Refrigeration and Heat Pumps. Part 2: Two-Phase Ejectors. *Inventions* **2019**, *4*, 16. [CrossRef]
10. Mil'man, O.O.; Anan'ev, P.A. Air-Cooled Condensing Units in Thermal Engineering (Review). *Therm. Eng.* **2020**, *67*, 872–891. [CrossRef]
11. Xu, H.; Jiang, S.; Xie, M.X.; Jia, T.; Dai, Y.J. Technical improvements and perspectives on humidification-dehumidification desalination—A review. *Desalination* **2022**, *541*, 116029. [CrossRef]
12. Narayan, G.P.; Sharqawy, M.H.; Summers, E.K.; Lienhard, J.H.; Zubair, S.M.; Antar, M.A. The potential of solar-driven humidification-dehumidification desalination for small-scale decentralized water production. *Renew. Sustain. Energy Rev.* **2010**, *14*, 1187–1201. [CrossRef]
13. Giwa, A.; Akther, N.; Al Housani, A.; Haris, S.; Hasan, S.W. Recent advances in humidification dehumidification (HDH) desalination processes: Improved designs and productivity. *Renew. Sustain. Energy Rev.* **2016**, *57*, 929–944. [CrossRef]
14. Al-Ismaili, A.M.; Jayasuriya, H. Seawater greenhouse in Oman: A sustainable technique for freshwater conservation and production. *Renew. Sustain. Energy Rev.* **2016**, *54*, 653–664. [CrossRef]
15. Kartik, S.; Balsora, H.; Sharma, M.; Saptoro, A.; Jain, R.K.; Joshi, J.B.; Sharma, A. Valorization of plastic wastes for production of fuels and value-added chemicals through pyrolysis—A review. *Therm. Sci. Eng. Prog.* **2022**, *32*, 101316. [CrossRef]
16. Qureshi, K.M.; Lup, A.N.K.; Khan, S.; Abnisa, F.; Daud, W.M.A.W. A technical review on semi-continuous and continuous pyrolysis process of biomass to bio-oil. *J. Anal. Appl. Pyrolysis* **2018**, *131*, 52–75. [CrossRef]
17. Wongwises, S.; Naphon, P. Heat-mass transfer and flow characteristics of two-phase countercurrent annular flow in a vertical pipe. *Int. Commun. Heat Mass Transf.* **1998**, *25*, 819–829. [CrossRef]
18. Han, H.; Gabriel, K. Flow physics of upward cocurrent gas-liquid annular flow in a vertical small diameter tube. *Microgravity Sci. Technol.* **2006**, *18*, 27–38. [CrossRef]
19. Ami, T.; Umekawa, H.; Ozawa, M. Dryout of counter-current two-phase flow in a vertical tube. *Int. J.Multiph. Flow* **2014**, *67*, 54–64. [CrossRef]
20. Abishek, S.; King, A.J.; Narayanaswamy, R. Computational analysis of—Two-phase flow and heat transfer in parallel and counter flow double-pipe evaporators. *Int. J. Heat Mass Transf.* **2017**, *104*, 615–626. [CrossRef]
21. Chen, W.B.; Tan, R.B.H. Theoretical analysis of two phase bubble formation in an immiscible Liquid. *AIChEJ.* **2003**, *49*, 1964–1971. [CrossRef]
22. Shilyaev, M.I.; Tolstykh, A.V. Simulation of heat and mass exchange in foam apparatus at high moisture content in vapor–gas mixture. *Theor. Found. Chem. Eng.* **2013**, *47*, 165–174. [CrossRef]
23. Lapteva, E.A.; Laptev, A.G. Models and calculations of the effectiveness of gas and liquid cooling in foam and film apparatuses. *Theor. Found. Chem. Eng.* **2016**, *50*, 430–438. [CrossRef]
24. Inaba, H.; Aoyama, S.; Haruki, N.; Horibe, A.; Nagayoshi, K. Heat and mass transfer characteristics of air bubbles and hot water by direct contact. *Heat Mass Transf.* **2002**, *38*, 449–457. [CrossRef]
25. Bezrodny, M.K.; Goliyad, N.N.; Barabash, P.A.; Kostyuk, A.P. Interphase heat-and-mass transfer in a flowing bubbling layer. *Therm. Eng.* **2012**, *59*, 479–484. [CrossRef]
26. Pianthong, K.; Seehanam, W.; Behnia, M.; Sriveerakul, T.; Aphornratana, S. Investigation and improvement of ejector refrigeration system using computational fluid dynamics technique. *Energy Convers. Manag.* **2007**, *48*, 2556–2564. [CrossRef]
27. Kim, H.D.; Setoguchi, T.; Yu, S.; Raghunathan, S. Navier-Stokes computations of the supersonic ejector-diffuser system with a second throat. *J. Therm. Sci.* **1999**, *8*, 79–83. [CrossRef]
28. Kou, J.; Li, Z. Numerical Simulation of New Axial Flow Gas-Liquid Separator. *Processes* **2021**, *10*, 64. [CrossRef]
29. Ji, L.; Zhao, Q.; Deng, H.; Zhang, L.; Deng, W. Experimental Study on a New Combined Gas–Liquid Separator. *Processes* **2022**, *10*, 1416. [CrossRef]

30. Farooqui, A.; Tomaso, F.D.; Bose, A.; Ferrero, D.; Llorca, J.; Santarelli, M. Techno-economic and exergy analysis of polygeneration plant for power and DME production with the integration of chemical looping CO_2/H_2O splitting. *Energy Convers.Manag.* **2019**, *186*, 200–219. [CrossRef]

31. Wotzka, A.; Jorabchi, M.N.; Wohlrab, S. Separation of H_2O/CO_2 mixtures by mfi membranes: Experiment and monte carlo study. *Membranes* **2021**, *11*, 439. [CrossRef]

32. Zong, X.; Liu, J.-P.; Yang, X.-P.; Yan, J.-J. Experimental study on the direct contact condensation of steam jet in subcooled water flow in a rectangular mix chamber. *Int. J. Heat Mass Trans.* **2015**, *80*, 448–457. [CrossRef]

33. Xu, Q.; Guo, L.; Zou, S.; Chen, J.; Zhang, X. Experimental study on direct contact condensation of stable steam jet in water flow in a vertical pipe. *Int. J. Heat Mass Transf.* **2013**, *66*, 808–817. [CrossRef]

34. Mahood, H.B.; Sharif, A.O.; Al-Aibi, S.; Hawkins, D.; Thorpe, R. Analytical solution and experimental measurements for temperature distribution prediction of three-phase direct-contact condenser. *Energy* **2014**, *67*, 538–547. [CrossRef]

35. Ma, X.; Xiao, X.; Jia, H.; Li, J.; Ji, Y.; Lian, Z.; Guo, Y.U. Experimental research on steam condensation in presence of non-condensable gas under high pressure. *Ann. Nucl. Energy* **2021**, *158*, 108282. [CrossRef]

36. Pedemonte, A.A.; Traverso, A.; Massardo, A.F. Experimental analysis of pressurised humidification tower for humid air gas turbine cycles. Part A: Experimental campaign. *Appl. Therm. Eng.* **2008**, *28*, 1711–1725. [CrossRef]

37. Pedemonte, A.A.; Traverso, A.; Massardo, A.F. Experimental analysis of pressurised humidification tower for humid air gas turbine cycles. Part B: Correlation of experimental data. *Appl. Therm. Eng.* **2008**, *28*, 1623–1629. [CrossRef]

38. Traverso, A. Humidification tower for humid air gas turbine cycles: Experimental analysis. *Energy* **2010**, *35*, 894–901. [CrossRef]

39. Caratozzolo, F.; Traverso, A.; Massardo, A.F. Implementation and experimental validation of a modeling tool for humid air turbine saturators. *Appl. Therm. Eng.* **2011**, *31*, 3580–3587. [CrossRef]

40. Zare, S.; Jamalkhoo, M.H.; Passandideh-Fard, M. Experimental Study of Direct Contact Condensation of Steam Jet in Water Flow in a Vertical Pipe With Square Cross Section. *Int. J. Multiph. Flow* **2018**, *1*, 74–88. [CrossRef]

41. Datta, P.; Chakravarty, A.; Ghosh, K. Experimental investigation on the effect of initial pressure conditions during steam-water direct contact condensation in a horizontal pipe geometry. *Int. Commun. Heat Mass Transf.* **2021**, *121*, 105082. [CrossRef]

42. Karapantsios, T.D.; Kostoglou, M.; Karabelas, A.J. Local condensation rates of steam-air mixtures in direct contact with a falling liquid film. *Int. J. Heat Mass Transf.* **1995**, *38*, 779–794. [CrossRef]

43. Karapantsios, T.D.; Karabelas, A.J. Direct-contact condensation in the presence of noncondensables over free-falling films with intermittent liquid feed. *Int. J. Heat Mass Transf.* **1995**, *38*, 795–805. [CrossRef]

44. Chen, X.; Tian, Z.; Guo, F.; Yu, X.; Huang, Q. Experimental investigation on direct-contact condensation of subatmospheric pressure steam in cocurrent flow packed tower. *Energy Sci. Eng.* **2022**, *10*, 2954–2969. [CrossRef]

45. Genić, S.B. Direct-contact condensation heat transfer on downcommerless trays for steam-water system. *Int. J. Heat Mass Transf.* **2006**, *49*, 1225–1230. [CrossRef]

46. Genić, S.B.; Jaćimović, B.M.; Vladić, L.A. Heat transfer rate of direct-contact condensation on baffle trays. *Int. J. Heat Mass Transf.* **2008**, *51*, 5772–5776. [CrossRef]

47. Chen, X.; Tian, M.; Zhang, G.; Leng, X.; Qiu, Y.; Zhang, J. Visualization study on direct contact condensation characteristics of sonic steam jet in subcooled water flow in a restricted channel. *Int. J. Heat Mass Transf.* **2019**, *145*, 118761. [CrossRef]

48. Fei, Y.; Xiao, Q.; Xu, J.; Pan, J.; Wang, S.; Wang, H.; Huang, J. A novel approach for measuring bubbles uniformity and mixing efficiency in a direct contact heat exchange. *Energy* **2015**, *93*, 2313–2320. [CrossRef]

49. Xu, J.; Xiao, Q.; Chen, Y.; Fei, Y.; Pan, J.; Wang, H. A modified L2-star discrepancy method for measuring mixing uniformity in a direct contact heat exchanger. *Int. J. Heat Mass Transf.* **2016**, *97*, 70–76. [CrossRef]

50. Mahood, H.B.; Thorpe, R.B.; Campbell, A.N.; Sharif, A.O. Experimental measurements and theoretical prediction for the transient characteristic of a two-phase two-component direct contact condenser. *Appl. Therm. Eng.* **2015**, *87*, 161–174. [CrossRef]

51. Mahood, H.B.; Campbell, A.N.; Thorpe, R.B.; Sharif, A.O. Heat transfer efficiency and capital cost evaluation of a three-phase direct contact heat exchanger for the utilisation of low-grade energy sources. *Energy Convers. Manag.* **2015**, *106*, 101–109. [CrossRef]

52. Mahood, H.B.; Sharif, A.; Thorpe, R.B. Transient volumetric heat transfer coefficient prediction of a three-phase direct contact condenser. *Heat Mass Trans.* **2015**, *52*, 165–170. [CrossRef]

53. Mahood, H.B.; Campbell, A.N.; Thorpe, R.B.; Sharif, A.O. Experimental measurements and theoretical prediction for the volumetric heat transfer coefficient of a three-phase direct contact condenser. *Int. Commun. Heat Mass Transf.* **2015**, *66*, 180–188. [CrossRef]

54. Mahood, H.B.; Campbell, A.N.; Sharif, A.O.; Thorpe, R.B. Heat transfer measurement in a three-phase direct-contact condenser under flooding conditions. *Appl. Therm. Eng.* **2016**, *95*, 106–114. [CrossRef]

55. Mahood, H.B.; Campbell, A.N.; Baqir, A.S.; Sharif, A.O.; Thorpe, R.B. Convective heat transfer measurements in a vapour-liquid-liquid three-phase direct contact heat exchanger. *Heat Mass Transf.* **2018**, *54*, 1697–1705. [CrossRef]

56. Mahood, H.B.; Campbell, A.N.; Thorpe, R.B.; Sharif, A.O. Measuring the overall volumetric heat transfer coefficient in a vapor-liquid–liquid three-phase direct contact heat exchanger. *Heat Trans. Eng.* **2018**, *39*, 208–216. [CrossRef]

57. Baqir, A.S.; Mahood, H.B.; Sayer, A.H. Temperature distribution measurements and modelling of a liquid-liquid-vapour spray column direct contact heat exchanger. *Appl. Therm. Eng.* **2018**, *139*, 542–551. [CrossRef]

58. Baqir, A.S.; Mahood, H.B.; Campbell, A.N.; Griffiths, A.J. Measuring the average volumetric heat transfer coefficient of a liquid–liquid–vapour direct contact heat exchanger. *Appl. Therm. Eng.* **2016**, *103*, 47–55. [CrossRef]
59. Baqir, A.S.; Mahood, H.B.; Hameed, M.S.; Campbell, A.N. Heat transfer measurement in a three-phase spray column direct contact heat exchanger for utilisation in energy recovery from low-grade sources. *Energy Convers. Manag.* **2016**, *126*, 342–351. [CrossRef]
60. Yang, X.P.; Liu, J.P.; Zong, X.; Chong, D.T.; Yan, J.J. Experimental study on the direct contact condensation of the steam jet in subcooled water flow in a rectangular channel: Flow patterns and flow field. *Int. J. Heat Fluid Flow* **2015**, *56*, 172–181. [CrossRef]
61. Pommerenck, J.; Alanazi, Y.; Gzik, T.; Vachkov, R.; Pmmerenck, J.; Hackleman, D.E. Recovery of a multicomponent, single phase aerosol with a difference in vapor pressures entrained in a large air flow. *J. Chem. Thermodyn.* **2012**, *46*, 109–115. [CrossRef]
62. Kwidzinski, R. Condensation heat and mass transfer in steam–water injectors. *Int. J. Heat MassTransf.* **2021**, *164*, 120582. [CrossRef]
63. Shah, A.; Chughtai, I.R.; Inayat, M.H. Experimental and numerical analysis of steam jet pump. *Int. J. Multiph. Flow* **2011**, *37*, 1305–1314. [CrossRef]
64. Shah, A.; Chughtai, I.R.; Inayat, M.H. Experimental study of the characteristics of steam jet pump and effect of mixing section length on direct-contact condensation. *Int. J. Heat Mass Trans.* **2013**, *58*, 62–69. [CrossRef]
65. Shah, A.; Chughtai, I.R.; Inayat, M.H. Experimental and numerical investigation of the effect of mixing section length on direct-contact condensation in steam jet pump. *Int. J. Heat Mass Trans.* **2014**, *72*, 430–439. [CrossRef]
66. Reddick, C.; Sorin, M.; Sapoundjiev, H.; Aidoun, Z. Effect of a mixture of carbon dioxide and steam on ejector performance: An experimental parametric investigation. *Exp. Therm. Fluid Sci.* **2018**, *92*, 353–365. [CrossRef]
67. Michalski, L.; Eckersdorf, K.; Kucharski, J.; McGhee, J. *Temperature Measurement*, 2nd ed.; John Wiley & Sons Ltd.: Hoboken, NJ, USA, 2001.
68. Lee, T.-W. *Thermal and Flow Measurements*; CRC Press: Boca Raton, FL, USA; Taylor & Francis Group: Abingdon, UK; LLC: Boca Raton, FL, USA, 2008.
69. Prosperetti, A.; Tryggvason, G. *Computational Methods for Multiphase Flow*; Cambridge University Press: Cambridge, UK, 2007.
70. Wang, J.; Chen, C.; Cai, Q.; Wang, C. Direct contact condensation of steam jet in subcooled water: A review. *Nucl. Eng. Des.* **2021**, *377*, 111142. [CrossRef]
71. Yi, K.; Kim, S.; Park, S. Numerical Study on Direct Contact Condensation Phenomenon of Saturated Steam. In Proceedings of the Transactions of the Korean Nuclear Society Autumn Meeting, Yeosu, Republic of Korea, 25–26 October 2018.
72. Colarossi, A.; Trask, N.; Schmidt, D.P.; Bergander, M.J. Multidimensional modeling of condensing two-phase ejector flow. *Int. J. Refrig.* **2012**, *35*, 290–299. [CrossRef]
73. Hughes, T.D. Multiphase CFD Analysis of Direct Contact Condensation Flow Regimes in a Large Water Pool. Master's Thesis, Texas A&M University, College Station, TX, USA, August 2019.
74. Wagner, W.; Kretzschmar, H.-J. *International Steam Tables, Properties of Water and Steam Based on the Industrial Formulation IAPWS-IF97: Tables, Algorithms, Diagrams, and CD-ROM Electronic Steam Tables: All of the Equations of IAPWS-IF97 Including a Complete Set of Supplementary Backward Equations for Fast Calculations of Heat Cycles, Boilers, and Steam Turbines*, 2nd ed.; Springer: Berlin/Heidelberg, Germany, 2002.
75. Thiele, R. Modeling of Direct Contact Condensation with OpenFoam. Master's Thesis, Division of Nuclear Technology, Royal Institute of Technology, Stockholm, Sweden, 2010.
76. Kunz, R.A. Preconditioned NavierStokes method for two-phase flows with application to cavitation prediction. *Comput. Fluid* **2000**, *29*, 849–875. [CrossRef]
77. Jayachandran, K.N.; Roy, A.; Ghosh, P. Numerical investigation on unstable direct contact condensation of cryogenic fluids. *IOP Conf. Ser. Mater. Sci. Eng.* **2017**, *171*, 012052. [CrossRef]
78. Anglart, H.; Nylund, O.; Kurul, N.; Podowski, M.Z. CFD prediction of flow and phase distribution in fuel assemblies with spacers. *Nucl. Eng. Des.* **1997**, *177*, 215–228. [CrossRef]
79. Sato, Y.; Sekoguchi, K. Liquid velocity distribution in two-phase bubble flow. *Int. J. Multiph. Flow* **1975**, *2*, 79–95. [CrossRef]
80. Ranz, W.; Marshall, W. Evaporation from Drops. *Chem. Eng. Prog.* **1952**, *48*, 141–146.
81. Schmidt, D.P.; Gopalakrishnan, S.; Jasak, H. Multidimensional simulation of thermal non-equilibrium channel flow. *Int. J. Multiph. Flow* **2010**, *36*, 284–292. [CrossRef]
82. Downar-Zapolski, P.; Bilicki, Z.; Bolle, L.; Franco, J. The non-equilibrium relaxation model for one-dimensional flashing liquid flow. *IJMF* **1996**, *22*, 473–483. [CrossRef]
83. Bergander, M.J.; Schmidt, D.P.; Wojciechowski, J.; Szklarz, M. Condensing Ejector for Second Step Compression in Refrigeration Cycles. In Proceedings of the International Refrigeration and Air Conditioning Conference, West Lafayette, IN, USA, 14–17 July 2008.
84. Zhang, Y.; Qu, X.; Zhang, G.; Leng, X.; Tian, M. Effect of non-condensable gas on the performance of steam-water ejector in a trigeneration system for hydrogen production: An experimental and numerical study. *Int. J. Hydrogen Energy* **2020**, *45*, 20266–20281. [CrossRef]
85. Shah, A.; Chughtai, I.R.; Inayat, M.H. Numerical simulation of Direct-contact Condensation from a Supersonic Steam Jet in Subcooled Water. *Chin. J. Chem. Eng.* **2010**, *18*, 577–587. [CrossRef]
86. Hughmark, G.A. Mass and heat transfer from a rigid sphere. *AICHE J.* **1967**, *13*, 1219–1221. [CrossRef]

87. Brucker, G.G.; Sparrow, E.M. Direct contact condensation of steam bubbles in water at high pressure. *Int. J. Heat Mass Transf.* **1977**, *20*, 371–381. [CrossRef]
88. Koirala, R.; Inthavong, K.; Date, A. Numerical study of flow and direct contact condensation of entrained vapor in water jet eductor. *Exp. Comput. Multiph. Flow* **2022**, *4*, 291–303. [CrossRef]
89. Takami, K.M.; Mahmoudi, J.; Time, R.W. A simulated H_2O/CO_2 Condeser Design for Oxy-fuel CO_2 Capture Process. *Energy Procedia* **2009**, *1*, 1143–1450. [CrossRef]
90. Hohne, T.; Gasiunas, S.; Seporaitis, M. Numerical Modelling of a Direct Contact Condensation Experiment. In Proceedings of the 2nd World Congress on Momentum, Heat and Mass Transfer (MHMT'17), Paper No. ICMFHT 102. Barcelona, Spain, 6–8 April 2017. [CrossRef]
91. Aya, I.; Nariai, H. Evaluation of heat-transfer coefficient at direct-contact condensation of cold water and steam. *Nucl. Eng. Des.* **1991**, *131*, 17–24. [CrossRef]
92. Cocci, R.; Ghione, A.; Sargentini, L.; Damblin, G.; Lucor, D. Model assessment for direct contact condensation induced by a sub-cooled water jet in a circular pipe. *Int. J. Heat Mass Transf.* **2022**, *195*, 123162. [CrossRef]

Article

Thermodynamic Analysis of Negative CO_2 Emission Power Plant Using Aspen Plus, Aspen Hysys, and Ebsilon Software

Paweł Ziółkowski [1], Paweł Madejski [2,*], Milad Amiri [1], Tomasz Kuś [2], Kamil Stasiak [1], Navaneethan Subramanian [2], Halina Pawlak-Kruczek [3], Janusz Badur [4], Łukasz Niedźwiecki [3] and Dariusz Mikielewicz [1]

[1] Faculty of Mechanical Engineering and Ship Technology, Institute of Energy, Gdańsk University of Technology, 80-233 Gdańsk, Poland; pawel.ziolkowski1@pg.edu.pl (P.Z.); milad.amiri@pg.edu.pl (M.A.); kamil.stasiak@pg.edu.pl (K.S.); dariusz.mikielewicz@pg.edu.pl (D.M.)

[2] Department of Power Systems and Environmental Protection Facilities, Faculty of Mechanical Engineering, AGH University of Science and Technology, 30-059 Kraków, Poland; kus@agh.edu.pl (T.K.); subraman@agh.edu.pl (N.S.)

[3] Department of Energy Conversion Engineering, Faculty of Mechanical and Power Engineering, Wroclaw University of Science and Technology, 50-370 Wrocław, Poland; halina.pawlak@pwr.edu.pl (H.P.-K.); lukasz.niedzwiecki@pwr.edu.pl (Ł.N.)

[4] Energy Conversion Department, Institute of Fluid Flow Machinery, Polish Academy of Sciences, 80-231 Gdańsk, Poland; jb@imp.gda.pl

[*] Correspondence: madejski@agh.edu.pl

Citation: Ziółkowski, P.; Madejski, P.; Amiri, M.; Kuś, T.; Stasiak, K.; Subramanian, N.; Pawlak-Kruczek, H.; Badur, J.; Niedźwiecki, Ł.; Mikielewicz, D. Thermodynamic Analysis of Negative CO_2 Emission Power Plant Using Aspen Plus, Aspen Hysys, and Ebsilon Software. *Energies* **2021**, *14*, 6304. https://doi.org/10.3390/en14196304

Academic Editors: Francesco Nocera, Kyung Chun Kim and Marco Marengo

Received: 18 July 2021
Accepted: 24 September 2021
Published: 2 October 2021

Abstract: The article presents results of thermodynamic analysis using a zero-dimensional mathematical models of a negative CO_2 emission power plant. The developed cycle of a negative CO_2 emission power plant allows the production of electricity using gasified sewage sludge as a main fuel. The negative emission can be achieved by the use this type of fuel which is already a "zero-emissive" energy source. Together with carbon capture installation, there is a possibility to decrease CO_2 emission below the "zero" level. Developed models of a novel gas cycle which use selected codes allow the prediction of basic parameters of thermodynamic cycles such as output power, efficiency, combustion composition, exhaust temperature, etc. The paper presents results of thermodynamic analysis of two novel cycles, called PDF0 and PFD1, by using different thermodynamic codes. A comparison of results obtained by three different codes offered the chance to verify results because the experimental data are currently not available. The comparison of predictions between three different software in the literature is something new, according to studies made by authors. For gross efficiency (54.74%, 55.18%, and 52.00%), there is a similar relationship for turbine power output (155.9 kW, 157.19 kW, and 148.16 kW). Additionally, the chemical energy rate of the fuel is taken into account, which ultimately results in higher efficiencies for flue gases with increased steam production. A similar trend is assessed for increased CO_2 in the flue gas. The developed precise models are particularly important for a carbon capture and storage (CCS) energy system, where relatively new devices mutually cooperate and their thermodynamic parameters affect those devices. Proposed software employs extended a gas–steam turbine cycle to determine the effect of cycle into environment. First of all, it should be stated that there is a slight influence of the software used on the results obtained, but the basic tendencies are the same, which makes it possible to analyze various types of thermodynamic cycles. Secondly, the possibility of a negative CO_2 emission power plant and the positive environmental impact of the proposed solution has been demonstrated, which is also a novelty in the area of thermodynamic cycles.

Keywords: CCS; CO_2 negative power plant; Aspen Plus; Aspen Hysys; Ebsilon

1. Introduction

Decarbonization of the economy, specifically in energy generation sector, has been adopted as a world-wide policy with signing of the Paris Agreement by nearly 200 signatories, including most significant emitters [1]. Thus, an ambitious greenhouse gases reduction goals has been set, in order to prevent the average global temperature increasing more than 1.5 °C above the pre-industrial levels [1]. An extensive effort is needed to achieve such goal [2]. Fossil fuels contributed approximately 9.5 Gt of carbon emitted to the atmosphere on average per year, as highlighted by the global carbon budget for years 2009–2018 [3].

The United Nations Framework Convention on Climate Change (UNFCCC) has recognized carbon capture and storage (CCS) technologies as important means of achieving ambitious climate goals [4]. Parameters, such as efficiency, cost, and water, have been considered as extremely important factors, determining the success of CCS technologies [5]. The work completed on CCS so far has been focused on post-combustion CCS [6], its integration with power plants [7,8], and combustion with different oxygen concentrations, since dilution of CO_2 in flue gases influences capturing efficiency [9,10]. Furthermore, various emerging CCS technologies, such as membrane-based carbon capture and storage [11], pre-combustion CO_2 capture [12], or carbon sequestration in hydrates [13–15], are also subjects of intensive investigations.

1.1. Concept of Negative Emissions Power Plants Using Biomass

The concept of achieving negative emissions has recently caught some attention [16]. Using biomass, combined with CCS, to achieve negative CO_2 emissions, it is often described as bioenergy with carbon capture and storage (BECSS) [17]. Investigative efforts have been mainly focused on chemical looping combustion (CLC) of biomass [18], as well as co-combustion with coal [19]. Lyngfelt et al. [20] investigated possibilities of leakages of stored CO_2 and concluded that, due to expected time scales of such events, the contribution of such leakages to the atmospheric stock would be relatively small, reaching approximately 3 ppm of CO_2 [20]. The use of different types of biomass has been investigated, including the work of Niu et al. [21] on CLC of sewage sludge. Saari et al. [22] investigated BECSS, using CLC with oxygen uncoupling dedicated to large scale co-generation plant. The results have shown an extremely small efficiency penalty of 0.7%, along with CO_2 capturing efficiency being as high as 97% [22].

Nonetheless, other ways to practically apply BECSS are also being investigated. Lisbona et al. [17] evaluated synergy between biogas plant and a biomass power plant, with special attention to the CCS module. Proposed installation, utilizing 1.5 MW of biomass and 1.4 MW of biogas (power as chemical energy at the inlet), was able to generate 750 kW$_{el}$ of electricity and generate 600 kW$_{th}$ of heat, for its own needs [17]. Additionally, the installation was able to capture 1620 tons of CO_2 per year [17]. Buscheck and Upadhye [23] investigated hybrid approach, incorporating oxy-combustion and heat accumulation. Such a concept is important, not only from the point of view of negative CO_2 emissions, but also from the point of view of limiting the curtailment of energy generation using intermittent renewable energy sources [23], as flexibility is critical for power systems with high shares of intermittent renewable energy sources (solar, wind) [24–28]. Capron et al. [29] focused on the use of Allam Cycle for achieving carbon negative emissions. A comprehensive overview, as presented in that paper, suggested that CCS could be combined with growing seafood, its subsequent processing, and production of biofuels, resulting in simultaneous increase in productivity and decrease in the exploited surface of the oceans, thus increasing the overall areas dedicated to conservation of biodiversity [29].

However, practical application of BECSS solution could be costly. Cheng et al. [30] determined levelized costs of different BECSS solutions for the US state of Virginia reaching USD/ton$_{CO_2}$ 82 (approx. EUR 70) for combustion of crop residues and USD/ton$_{CO_2}$ 137 (approx. EUR 115) for combustion of woody residues. This is still much less than the current market value that could be assigned for a ton of avoided CO_2 emissions [31]. However, a study performed by Restrepo-Valencia and Walter [32] indicates that EUR/ton$_{CO_2}$ 59 can be achieved for optimized BECSS using bagasse and the cost could be further decreased to EUR/ton$_{CO_2}$ 48 for larger plants. This suggests that significant amount of work is needed to optimize BECSS in terms of CAPEX and OPEX. Such goal can be achieved by optimization of such systems, by comprehensive thermodynamic analysis.

1.2. Software for Zero-Dimensional Modelling

The zero-dimensional approach is mainly used for systems optimization. A limited amount of the obtained data makes it possible to conduct many optimizing calculations of the turbine parameters or entire complex system composed of many devices, such as compressors, expanders, heat exchangers, combustion chambers, reactors, fuel cells, pumps, or ejectors.

Literature on different software is very extensive; however, the most widely used ones are presented below, as follows:

- Aspen Plus is intended for a combined system, steam cycle, ORC cycle; operation under 50–110% nominal load [33];
- Aspen Hysys is intended for a combined system; operation under 50–110% nominal load and dynamic conditions [34];
- Ebsilon is designed for advanced steam block systems and combined systems, operation under variable conditions 40–120% of nominal load [35,36];
- Gate Cycle is designed for advanced combination systems, variable load operation 40–120% of nominal load [37];
- COM-GAS is intended for design level of combined systems with full analysis of a heat recovery steam generator, pulverized fuel, and fluidized bed boilers [38,39];
- DIAGAR is intended for design and diagnostic level of steam systems with full steam turbine analysis [40];
- IPSEpro is a process simulation tool, which is equation-oriented and has been used for power plant simulations, including modeling of chemical looping CCS systems [22];

The most important issue about software for thermodynamic cycles is that they have a high degree of certainty and confidence in the calculation results, which are only achieved by highly validated codes. This means that such codes, in addition to basic calculation algorithms, have extensive expert procedures for checking the results before they are passed on to the user. We selected three codes for detailed analysis of the considered case, namely Aspen Plus, Aspen Hysys, and Ebsilon. The following subsections provide a literature review on these codes.

1.3. Scope and Aim

The main objective of this paper is to analyze an innovative technology together with the proof of concept, confirming the possibility of the use of sewage sludge to produce electricity while having a positive impact on the environment. The synergy between the CCS plant and the proposed utilization of sewage sludge (which is considered a renewable energy source) enables the installation to achieve overall negative emissions of CO_2 (nCO2PP). Proposed processes of utilization (PFD0—Sections 2 and 3; PFD1—Sections 4 and 5), called nCO2PP (negative CO_2 Power Plant), ensures reaching of scientific objectives related to three essential theoretical elements, namely: (1) a system that processes sewage sludge into syngas; (2) a system that burns the resulting fuel in pure oxygen in a dedicated wet combustion chamber; and (3) a system of a unique turbine cooperating with a spray ejector condenser with carbon dioxide capture.

The second aim of the article is to compare the results obtained in three computing codes, namely Aspen Plus, Aspen HYSYS, and Ebsilon, based on the assumption presented in next section, and subsequently pointing out the differences and identifying the reasons for them. Section 2 examines the original simple system consisting of an arrangement of equipment such as compressors, expanders, heat exchanger, combustion chamber, pump, and generator to generate electrical energy. A schematic of the cycle can be found in Figure 1, while Figure 2 presents the model in Aspen Hysys, Figure 3 in Aspen plus, and Figure 4 in Ebsilon. Section 3 presents the following subsections as follows: (1) thermodynamic parameters and mass flow rates in nodal points; (2) the output and efficiencies of power; and (3) the effect of NO_x production on combustion chamber temperature. In Section 4, this system is extended to include a spray ejector condenser, where diagrams of power output, efficiency, and chemical energy flow delivered to the combustion chamber are prepared for clarity of results. In Section 5, it is shown that this gas-fired power plant, after the use of gasification fuel (the composition of mixture 1 is given as an example), is CO_2-negative. The last section summarizes the work carried out and draws conclusions.

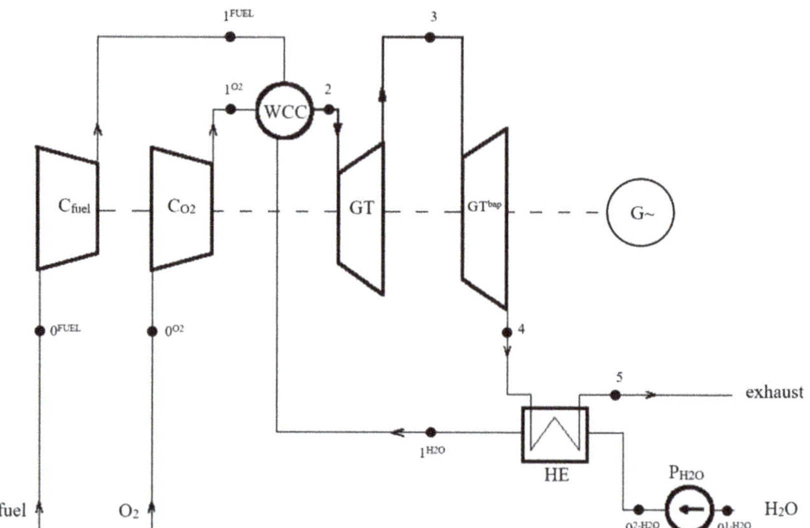

Figure 1. Process flow diagram of a gas mixture cycle PFD0—a steam-gas turbine system (0FUEL, 0O$_2$, 01-H$_2$O, 02-H$_2$O, 1FUEL, 1O$_2$, 1H$_2$O, 2, 3, 4, 5—cycle nodal points).

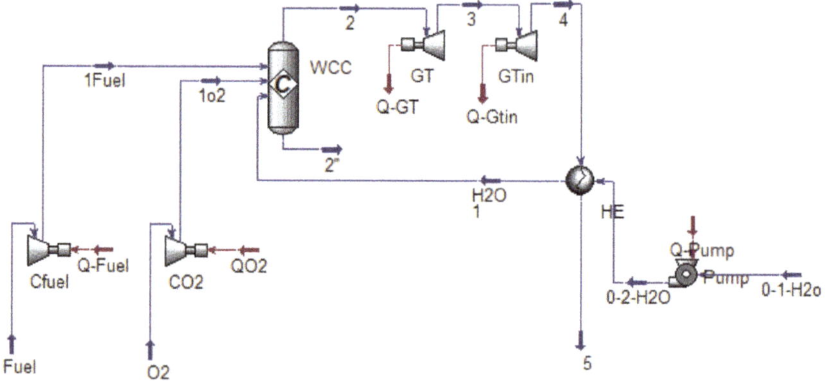

Figure 2. Simulation of PFD0 by Aspen Hysys (0FUEL, 0O$_2$, 01-H$_2$O, 02-H$_2$O, 1FUEL, 1O$_2$, 1H$_2$O, 2, 3, 4, 5—cycle nodal points).

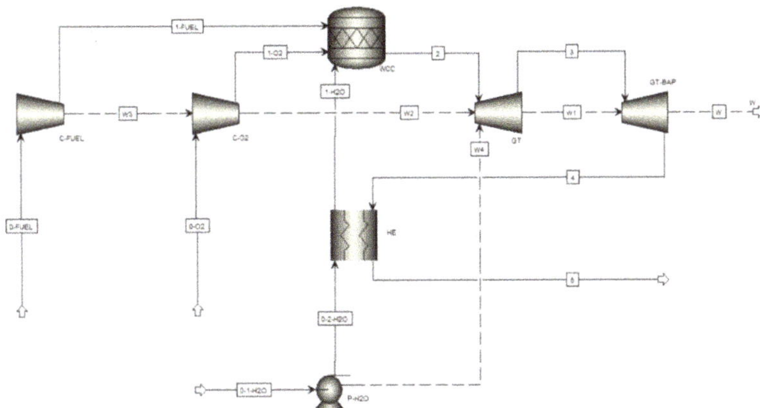

Figure 3. Simulation of PFD0 by Aspen Plus (0FUEL, 0O$_2$, 01-H$_2$O, 02-H$_2$O, 1FUEL, 1O$_2$, 1H$_2$O, 2, 3, 4, 5—cycle nodal points).

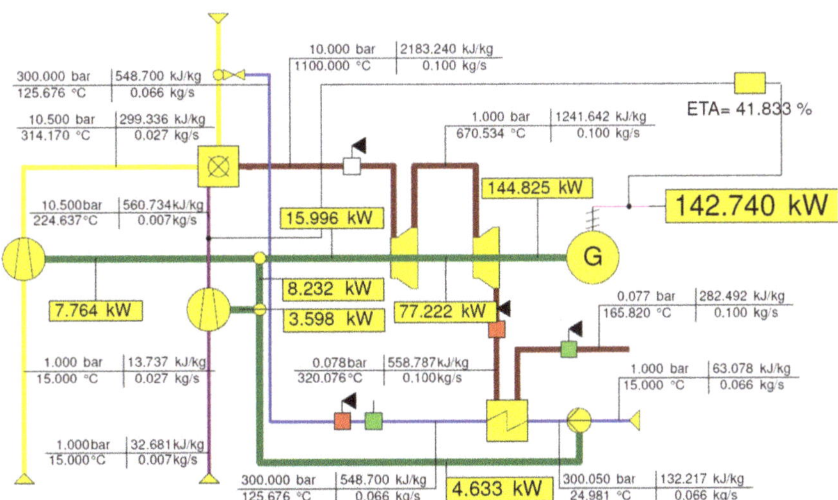

Figure 4. Simulation of PFD0 by Ebsilon (0FUEL, 0O$_2$, 01-H$_2$O, 02-H$_2$O, 1FUEL, 1O$_2$, 1H$_2$O, 2, 3, 4, 5—cycle nodal points).

2. Thermodynamic Cycle Considered in Three Software

2.1. Modeling and Simulation of Thermodynamic Cycles

The use of thermodynamic simulation software can strongly support designing, monitoring, and optimizing CCUS processes as the new solutions for existing and planned to build power plants.

Different perspectives of modeling has been created by Aspen Plus, such as steam power plant [41], predicting emissions of NO and N$_2$O from coal combustion [42], catalytic coal gasification infixed beds [43], biomass gasification in fluidized bed reactor [44], and in combined heat and power (CHP) biomass bubbling fluidized bed gasification unit coupled with an internal combustion engine (ICE) [45]. Ebsilon®Professional is a simulation software designed for performing simulations of processes in thermodynamic cycles, as well as steady-state and quasi dynamic simulations [46–49]. The Ebsilon library has an extensive number of components, useful for efficient calculations [49]. By placing the components in the system, a system of equations is generated based on mass and energy balance depending upon the component, which is solved by the Gauss–Seidel method. The iteration ends when the convergence criterion of 10-9 is reached for pressure, flow, and enthalpy variables [50]. Aspen HYSYS is defined as an industry-leading process modeling tool for conceptual study, strategic planning, management of asset, maximization and operational testing for gas processing, petroleum refining, oil and gas production, and air separation industries. Although HYSYS is mainly useful for oil and gas process industry, it is developed for various industries as follows [51]: ethanol plant; petroleum industry; heavy chemical industry; natural gas process plant; petrochemical industry; synthesis gas production; acid gas sweetening with DEA (Diethanolamine); biodiesel plant, etc. A comparison of units in Aspen Plus, Aspen HYSYS, and EBSILON is presented in Table 1.

Table 1. Comparison of units in Aspen Plus, Aspen HYSYS, and EBSILON.

Unit Operation	Aspen Plus	Aspen HYSYS	EBSILON
Stream mixing	Mixer	Mixer	Simple mixer
Component splitter	Sep, Sep2	Component Splitter	Simple splitter
Decanter	Decanter	3-Phase Separator	Selective splitter
Piping	Pipe, Pipeline	Pipe Segment, Compressible Gas Pipe	Pipe
Valves and fittings	Valve	Valve, Tee, Relief Valve	Valve
Equilibrium reactor	REquil	Equilibrium Reactor	Combustion chamber
Gibbs reactor	RGibbs	Gibbs Reactor	Gibbs reactor
Heat exchanger	HeatX, HxFlux, Hetran, HTRI-Xist	Heat Exchanger	Heat exchanger
Compressor	Compr, MCompr	Compressor	Compressor
Turbine	Compr, MCompr	Expander	Gas expander
Pump	Pump	Pump	Pump

Differences and similarities of thermodynamic parameters for the three used software including Aspen Hysys, Plus, and Ebsilon are indicated in Table 2. Crucial parameters for thermodynamic is its efficiency, which depend from many issues, but one of the important is model of fluid. The net system efficiency of the system was calculated according to the formula:

$$\eta_{net} = \frac{N_t - N_{C-fuel} - N_{C-O_2} - N_{P-H_2O} - N_{CCU} - N_{p-SEC}}{\dot{Q}_{CC}} \tag{1}$$

where:

N_t—combined turbine power on the shaft in [kW],

N_{C-fuel}—power for fuel compressor in [kW],

N_{C-O_2}—power for oxygen compressor in [kW],

N_{P-H_2O}—power for water pump PH$_2$O in [kW],

N_{P-SEC}—power for water pump PSEC supplying SEC in [kW],

N_{CCU}—combined power for CO$_2$ capture unit compressors [kW],

$\dot{Q}_{CC}$—chemical energy rate of combustion in [kW].

Important is also power for own needs as a sum:

$$N_{CP} = N_{C-fuel} + N_{C-O_2} + N_{P-H_2O} + N_{P-SEC} + N_{CCU} \tag{2}$$

According this equation, the powers depend from thermodynamic model of fluid which is possible to describe the real gas equation in a more precise form, which takes the form of the Peng–Robinson gas model:

$$p = \frac{\tilde{R}T}{v_M - b} - \frac{a\alpha_m}{v_M^2 + 2bv_M - b^2}, \tag{3}$$

where: $\tilde{R}$—universal gas constant, v_M—molar volume and

$$a = \frac{0.4572\tilde{R}^2 T^2_{cr(m)}}{p_{cr(m)}} \tag{4}$$

where: $p_{cr(m)}$—critical pressure, $T_{cr(m)}$—critical temperature. Another constant is:

$$b = \frac{0.0778\tilde{R}T_{cr(m)}}{p_{cr(m)}} \tag{5}$$

and the last constant from the Formula (3) is expressed as:

$$\alpha_m = \left(1 + \xi_m\left(1 - T^{0.5}_{r(m)}\right)\right)^2 \tag{6}$$

assuming that the reduced temperature $T_{r(m)}$ expresses the ratio:

$$T_{r(m)} = \frac{T}{T_{cr(m)}} \tag{7}$$

and

$$\xi_m = 0.37464 + 1.54226\omega_m - 0.26992\omega_m^2 \tag{8}$$

where ω_m is the material constant expressing the molecular non-sphericity (centrality) of the particles. For example, for noble gases such as argon, krypton, neon, and xenon $\omega_m = 0$. It should be also mentioned that ω_m is determined for $T_{r(m)} = 0.7$ and can be determined by the relationship:

$$\omega_m = -\log_{10}\left(p^{sat}_{r(m)}\right) - 1 \tag{9}$$

where $p^{sat}_{r(m)}$ is the reduced evaporation pressure expressed as the relationship:

$$p^{sat}_{r(m)} = \frac{p_{sat(m)}}{p_{cr(m)}} \tag{10}$$

where $p_{sat(m)}$ is the saturation pressure (evaporation) for $T_{r(m)} = 0.7$.

Although Peng-Robinson as a thermodynamic model is used for both Aspen Hysys and Plus, thermodynamic tables for steam and Peng-Robinson for another working fluid are used in Ebsilon.

Table 2. Differences and similarities for calculations.

Parameter	Symbol	Unit	
Thermodynamic model	Peng-Robinson	-	Thermodynamics tables for steam and Peng-Robinson for another working fluid
Net efficiency	η_net	-	$\dfrac{N_t - N_{C-fuel} - N_{C-O_2} - N_{P-H_2O} - N_{CCU} - N_{P-SEC}}{\dot{Q}_{CC}}$
Gross efficiency	η_g	-	$\eta_g = \dfrac{N_t}{\dot{Q}_{CC}}$
NO$_x$ production	NO and NO$_2$	-	Without NO$_x$ production calculation in Ebsilon software
Chemical energy rate	$\dot{Q}_{CC}$	kW	$\dot{Q}_{CC} = \dot{m}_{fuel}LHV$
Reactions	combustion	-	Defined and could be modified

It should be underscored that the specific enthalpy of the fluid $h = h(p;T;Y_{(k)})$ is determined at the characteristic points by the thermodynamic table and depends on thermodynamic parameters, such as temperature T, pressure p, and specific components within the mixture of air and exhaust gases $Y_{(k)}; k = N_2, \ldots, Ar$ [52,53]. Another difference can be tangible in NO$_x$ production so that Aspen Hysys and Plus calculate NO$_x$ production including NO and NO_2, whilst it is not estimated in Ebsilon. In addition, reactions used in wet combustion chamber need to be defined in properties tab (Reaction's part) in Aspen Hysys. As it can be vividly seen, the method of calculating net efficiency, gross efficiency, and chemical energy rate is the same for three used software.

2.2. Thermodynamic Cycle

The thermodynamic cycle of the gas–steam turbine system is represented in Figure 1. The gas–steam turbine system consists of two gas–steam expanders, i.e., the gas–steam turbine (GT) part and the low-pressure gas–steam turbine below ambient pressure (GT_{bap}) with power generators (G~), the fuel compressor (C_{fuel}), the oxygen compressor (C_{O_2}), the water pump (P_{H_2O}), the heat exchanger (HE) for regenerative water heating, and the wet combustion chamber (WCC). The working fluid in the cycle is the gas–steam—a mixture of water vapor (H_2O) and carbon dioxide (CO_2). As observed in Figure 1, after increasing the pressure of selected fuel (methane and mixture 1) and O_2 in their related compressor, they are fed to a wet combustion chamber. Wet combustion chamber combusts selected fuels in the presence of oxygen O_2 to produce hot steam and carbon dioxide. Using the recycled water leads hot steam and carbon dioxide to cool within the wet combustion chamber to the desired temperature of a gas turbine. GT and GT_{bap} are used to decrease high-pressure (10 bar) working fluid (water vapor and carbon dioxide) to below ambient pressure (0.078 bar). A heat exchanger is not only simulated to achieve the cooled steam but also increases the temperature of water.

2.3. Assumptions for Cycle Modeling

Assumptions for the thermodynamic cycle, internal efficiency, and mechanical efficiency are illustrated in Tables 3–5. It can be noticed that the temperature of exhaust gas after WCC (before GT) is 1100 °C in Aspen Hysys and Plus and Ebsilon for (Ebsilon t_2 = const), while for Ebsilon t_2 = var is 1073 °C for methane and 1091 °C for mixture 1, respectively. These temperatures (namely 1100 °C in Aspen Plus and Aspen Hysys, and 1091 °C and 1073 °C) in front of the turbine were achieved by assuming a constant temperature of water feeding the combustion chamber, namely $t_1 H_2O$ = const = 125.1 °C. In addition, when the exhaust temperature after WCC is constant (t_2 = 1100 °C), water temperature before the combustion chamber is variable, respectively, 149.02 °C in Ebsilon with mixture 1, 131.84 °C in Ebsilon with methane, and 125.1 °C for both Aspen Hysys and Plus. Heat efficiency of the combustion chamber in Aspen Plus and Aspen Hysys is 99.9%. The rest of the assumptions for the three used software is the same.

Table 3. Assumptions for the thermodynamic cycle calculation using Aspen HYSYS, Aspen Plus, and Ebsilon.

Parameters	Symbol	Unit	Value
Mass flow of exhaust gas at the outlet from combustion chamber WCC	m_2	g/s	100
Air-fuel ratio in WCC	λ	-	1 (stoichiometric)
Pressure before GT	p_2	bar	10
Pressure after GT	p_3	bar	1
Pressure after GT_{bap}	p_4	bar	0.078
Water pressure to WCC	p_{1-H_2O}	bar	300
Temperature exhaust after WCC (before GT)	t_2	°C	1100 (1100 and variable in Ebsilon)
Initial water temperature (before PH2O pump)	t_{0-1-H_2O}	°C	15
Initial fuel temperature	t_{fuel}	°C	15
Initial oxygen temperature	t_{O_2}	°C	15
Initial fuel pressure (before Cfuel compressor)	p_{0-fuel}	bar	1
Initial oxygen pressure (before CO_2 compressor)	p_{0-O_2}	bar	1
Fuel to WCC pressure loss factor	δ_{fuel}	-	0.05
Oxygen to WCC pressure loss factor	δ_{O_2}	-	0.05
Oxygen purity		%	100

Table 3. *Cont.*

Parameters		Symbol	Unit	Value
Fuel mass flow	methane	$\dot{m}_{fuel}$	g/s	6.72
	Mixture—syngas	$\dot{m}_{fuel}$	g/s	18.00
Temperature exhaust after WCC (before GT)	Variable temperature in point $1H_2O$ (118.45; 131.84 and 125.1 °C)	$t_2 = const$	°C	1100
	Constant temperature in point $1H_2O$ (125.1 °C)	$t_2 = var$	°C	1100 1073 for mixture, 1091 for methane in Ebsilon
CO_2 fraction from combustion of methane	Methane	$X_{CO_2} = const$	mol%	8.47
	Mixture	$X_{CO_2} = var$	mol%	11.75 11.73 in Ebsilon
Water temperature before combustion chamber	Variable temperature exhaust after WCC	$t_{1\ H_2O} = const$	°C	125.1
	Constant temperature exhaust after WCC	$t_{1\ H_2O} = var$	°C	149.02 for mixture and 131.84 for methane in Ebsilon

Table 4. Assumed internal efficiency (adiabatic for Hysys and isentropic for Aspen Plus and Ebsilon).

Internal Efficiency	Symbol	Unit	Value
Turbine GT	η_{iGT}	-	0.89
Turbine GT$_{bap}$	$\eta_{iGT\text{-}bap}$	-	0.89
Fuel compressor C$_{fuel}$	$\eta_{iC\text{-}fuel}$	-	0.87
Oxygen compressor C$_{O_2}$	$\eta_{iC\text{-}O_2}$	-	0.87
Water pump P$_{H_2O}$	$\eta_{iP\text{-}H_2O}$	-	0.43

Table 5. Assumed mechanical efficiency—for Aspen Hysys it is impossible to change value.

Internal Efficiency	Symbol	Unit	Aspen HYSYS	Aspen Plus/EBSILON
Turbine GT	η_{mGT}	-	1	0.99
Turbine GT$_{bap}$	$\eta_{mGT\text{-}bap}$	-	1	0.99
Fuel compressor C$_{fuel}$	$\eta_{mC\text{-}fuel}$	-	1	0.99
Oxygen compressor C$_{O_2}$	$\eta_{mC\text{-}O_2}$	-	1	0.99
Water pump P$_{H_2O}$	$\eta_{mP\text{-}H_2O}$	-	1	0.99

2.4. Fuels

Syngas fuels produced from gasification are expected to be of different compositions, mainly due to inherent variability of sewage sludge composition, as reported by Werle and Wilk [54]. Therefore, two types of fuel were selected for the analysis, and compositions are presented in Figure 5. The first one is the syngas mixture which contains CO ($9.09\%_{mol}$); CO_2 ($25.61\%_{mol}$); CH_4 ($13.64\%_{mol}$); C_3H_8 ($3.39\%_{mol}$); H_2 ($45.16\%_{mol}$); and NH_3 ($3.10\%_{mol}$). However, the mass fractions of species for Aspen Plus and Aspen Hysys was introduced as data, namely CO ($13.31\%_{mass}$); CO_2 ($59.31\%_{mass}$); CH_4 ($11.46\%_{mass}$); C_3H_8 ($8.03\%_{mass}$); H_2 ($5.10\%_{mass}$); and NH_3 ($2.79\%_{mass}$). Selected compositions of the producer gas are well within the ranges of values are reported by Achweizer et al. [55] or Akkache et al. [56]. Methane fuel is added for comparison purposes. The compositions of selected fuels, including methane and mixture (syngas), are shown in Figure 5.

The values of LHV for mixture 1 and methane at 15 °C and 1 atm are presented in Table 6. It is noteworthy that Ebsilon uses empirical formulae based on elementary analysis, whereas LHV used for both Aspen Hysys and Plus are the same. Syngas is produced by gasifying sewage sludge.

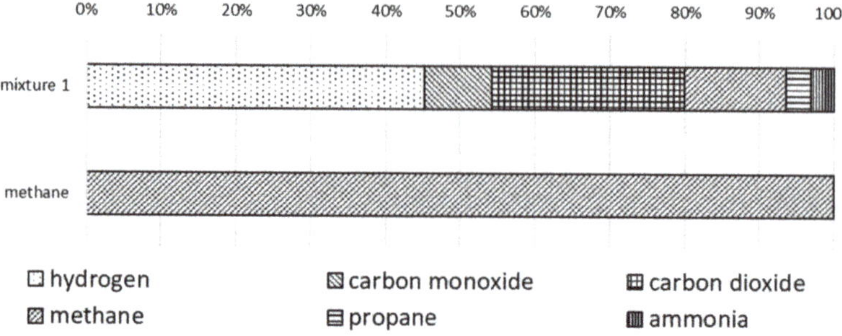

Figure 5. Fuel compositions for the analysed cycle.

Table 6. LHV based on ISO 6976:1995(E) for gas mixtures, value at 15 °C and 1 atm derived from Aspen and Ebsilon.

Software	LHV, MJ/kg	
	Syngas—Mixture	Methane
Aspen HYSYS and Aspen PLUS	17.079	50.035
Ebsilon	17.081	50.015

3. Results and Comparison

The most important nodal point results are presented in Section 3.1, while Section 3.2 refers to the efficiency results and Section 3.3 deals with the combustion of ammonia to various nitrogen compounds.

3.1. Nodal Points

Cycle nodal points for mixture (syngas) and methane are depicted in Tables 7 and 8, respectively. Having studied the data from Table 7, it can be considered that mass flow of mixture (syngas), O_2, and H_2O are 18 (g/s), 23.19 (g/s), and 58.80 (g/s) in Aspen Hysys and Plus, whilst these values in Ebsilon are 18 (g/s), 22.84 (g/s), and 59.164 (g/s), respectively. It is noticeable that simulation in Ebsilon was performed for two values of t_2 (temperature after WCC), as mentioned in Table 3. The temperature after compressor of fuel (syngas) is 255.6 °C, 253.33 °C, and 252.38 °C and after compressor of O_2 is 314.8 °C, 315.08 °C, and 314.17 °C in Aspen Hysys, Plus, and Ebsilon, respectively. Other differences can be observed in temperature before and after the heat exchanger. More accurately, temperature before heat exchanger is 25.11 °C for both simulation in Apen Hysys and Plus, whereas its value is 24.98 °C in Ebsilon. These temperatures are obtained by increasing the pressure in the pump.

As the same way, the temperature after heat exchanger is 125.11 °C was the same for both simulations in Apen Hysys and Plus, while it is 125.11 °C and 149.02 °C for mixture 1, 125.11 °C and 131.84 °C for methane when t_2 = var and t_2 = const in Ebsilon, respectively. In addition, CO_2, H_2O, and NO (N_2 in Ebsilon) result from combustion in a wet combustion chamber. As seen in Table 7, mole fraction of CO_2 is 11.75 and 11.73 in Aspen Hysys, Plus, and Ebsilon, respectively. Moreover, mole fraction of H_2O is 87.63 and 87.98 for mentioned software, respectively. The most important difference in arising composition is in the type of NO_x, so that there is NO (0.62) in Aspen Hysys and Plus, whereas N_2 (0.32) is created in Ebsilon. Although it is assumed that the temperature after wet combustion chamber is 1100 °C for Aspen Hysys, Plus, and Ebsilon, a different temperature (1073 °C) after WCC was simulated in Ebsilon. Results show that the maximum temperature of exhaust gases after the heat exchanger results from simulation of Aspen Plus and its value is 183.58 °C whereas the minimum one (147.3 °C) belongs to the simulation using Ebsilon (t_2 = const).

Table 7. Cycle nodal points on basis of syngas—mixture of gases as a fuel.

Parameter	Case	Unit	0^{Fuel}	1^{Fuel}	0^{O_2}	1^{O_2}	$0^{1\text{-}H_2O}$	$0^{2\text{-}H_2O}$	1^{H_2O}	2	3	4	5
Node Designation	-	-											
Mass flow $\dot{m}$	Aspen Hysys	g/s			23.2	23.2	58.8	58.8	58.8				
	Aspen Plus		18.0	18.0						100	100	100	100
	Ebsilon t_2 = var				22.4	22.4	59.6	59.6	59.6				
	Ebsilon t_2 = const												
O_2 fraction (X_{O_2})	Aspen Hysys	mol%	-	-			-	-	-				
	Aspen Plus		-	-	100	100	-	-	-	0.00	0.00	0.00	0.00
	Ebsilon t_2 = var		-	-			-	-	-				
	Ebsilon t_2 = const		-	-			-	-	-				
CO_2 fraction (X_{CO_2})	Aspen Hysys	mol%	-	-	-	-	-	-	-	11.75	11.75	11.75	11.75
	Aspen Plus		-	-	-	-	-	-	-				
	Ebsilon t_2 = var		-	-	-	-	-	-	-	11.73	11.73	11.73	11.73
	Ebsilon t_2 = const		-	-	-	-	-	-	-				
H_2O fraction (X_{H_2O})	Aspen Hysys	mol%	-	-	-	-				87.63	87.63	87.63	87.63
	Aspen Plus		-	-	-	-	100	100	100				
	Ebsilon t_2 = var		-	-	-	-				87.96	87.96	87.96	87.96
	Ebsilon t_2 = const		-	-	-	-							
NO fraction (N_2 in Ebsilon) (X_{NO})	Aspen Hysys	mol%	-	-	-	-	-	-	-				
	Aspen Plus		-	-	-	-	-	-	-	0.62	0.62	0.62	0.62
	Ebsilon t_2 = var		-	-	-	-	-	-	-				
	Ebsilon t_2 = const		-	-	-	-	-	-	-	0.31	0.31	0.31	0.31
Temperature (t)	Aspen Hysys	°C		255.6		314.8		25.11	125.11		672.5	324.7	178.6
	Aspen Plus		15	253.33	15	315.08	15			1100	672.51	323.64	183.58
	Ebsilon t_2 = const			252.38		314.17			149.02		673.58	324.86	147.3
	Ebsilon t_2 = var							24.98	125.11	1073	652.98	310.38	167.64
Pressure (p)	Aspen Hysys	bar											
	Aspen Plus		1	10.5	1	10.5	1	300	300	10	1	0.078	0.078
	Ebsilon t_2 = var												0.077
	Ebsilon t_2 = const												

Table 8. Cycle nodal points on basis of methane.

Parameter	Symbol	Unit	0^{Fuel}	1^{Fuel}	0^{O_2}	1^{O_2}	$0^{1\text{-}H_2O}$	$0^{2\text{-}H_2O}$	1^{H_2O}	2	3	4	5
Node Designation	-	-											
Mass flow $(\dot{m})$	Aspen Hysys	g/s											
	Aspen Plus		6.72	6.72	26.80	26.80	66.48	66.48	66.48	100	100	100	100
	Ebsilon t_2 = var												
	Ebsilon $t_2.$ = const												
O_2 fraction (X_{O_2})	Aspen Hysys	mol%											
	Aspen Plus		-	-	100	100	-	-	-	0.00	0.00	0.00	0.00
	Ebsilon t_2 = var												
	Ebsilon t_2 = const												
CO_2 fraction (X_{CO_2})	Aspen Hysys	mol%											
	Aspen Plus		-	-	-	-	-	-	-	8.47	8.47	8.47	8.47
	Ebsilon t_2 = var												
	Ebsilon t_2 = const												
H_2O fraction (X_{H_2O})	Aspen Hysys	mol%									91.53	91.53	91.53
	Aspen Plus		-	-	-	-	100	100	100	91.53			
	Ebsilon t_2 = var												
	Ebsilon t_2 = const												
Temperature (t)	Aspen Hysys	°C		225.39		314.8			125.11		667.3	318.4	158.6
	Aspen Plus		15		15	315.08	15	25.11		1100	669.51	318.99	165.82
	Ebsilon t_2 = const			224.63		314.17		24.98	131.84		670.49	320.01	155.65
	Ebsilon t_2 = var								125.11	1091	663.9	315.35	161.47
Pressure (p)	Aspen Hysys	bar											
	Aspen Plus		1	10.5	1	10.5	1	300	300	10	1	0.078	0.078
	Ebsilon t_2 = var												
	Ebsilon t_2 = const												0.077

As it can be observed that the difference in temperature (324.7 °C in Hysys, 323.64 °C in Plus, and 324.82 °C in Ebsilon) before heat exchanger for gas–steam is less than 0.4%, the type of heat exchanger plays an indispensable role in regard to the value of cooled gas–steam (exhaust gases). In addition, pressure drop of heat exchanger is zero in Aspen Hysys and Plus, but a pressure drop is not constant in Ebsilon (pressure differences between point 4 and 5). Moreover, decreasing the temperature after wet combustion chamber leads to increasing the temperature of cooled gas–steam (t_5), so that approximately a 12% increase in temperature of steam after heat exchanger results from decreasing the temperature after WCC from 1100 to 1073 °C.

Cycle nodal points for methane are indicated in Table 8. Mass flow of fuel, O_2, and H_2O are the same for used software and its value is 6.72 (g/s), 26.80 (g/s), and 66.48 (g/s), respectively. Although the temperature of water that was fed to WCC is 125.11 °C, this value indicates for simulation various temperature in combustion chamber ($t_2 = 1091$ °C) in Ebsilon and ($t_2 = 1100$ °C) in Aspen Plus and Aspen Hysys. To obtain the same temperature in combustion chamber in simulation using a Ebsilon, we need to increase temperature to 131.84 °C. Moreover, a decrease (0.82%) in the temperature of exhaust gases after WCC from 1100 °C to 1091 °C results in an increase (3.6%) in temperature of cooled gas–steam after heat exchanger from 155.65 °C to 161.47 °C in Ebsilon.

As a result, changing the type of fuels leads to a change in the compositions of exhaust gases and temperature after a heat exchanger. For example, approximately 88% and 92% mole fraction of H_2O result from mixture 1 (syngas) and methane, respectively. Furthermore, using a mixture of gases as a fuel and methane result in creating approximately 12% and 8% mole fraction of CO_2. Moreover, the average temperature (among three software) after the heat exchanger is 169.3 °C for mixture and 160 °C for methane.

3.2. Efficiency and Summarized Effects

Summarized results for two fuels (mixture 1 and methane) in three used software including Aspen Hysys, Plus, and Ebsilon are illustrated in Table 9. The mass flow rate after WCC is 100 g/s for the three mentioned software. The gross power of turbines for mixture is 154.37 kW, 154.20 kW, 154.72 kW ($t_2 = $ const), and 151.36 kW ($t_2 = $ var) in Aspen Hysys, Plus, and Ebsilon, respectively. It can be observed that less than a 0.34% difference was obtained among three software, when t_2 is 1073 °C or 1100 °C. On the other hand, for methane, these values are 161.42 kW, 160.72 kW, 160.89 kW ($t_2 = $ const), and 159.76 kW ($t_2 = $ var) for the mentioned software, respectively. The results show that, at the same assumption, changing the type of fuels from mixture 1 to methane leads the gross power output of turbines to increase approximately by 4% in Aspen Hysys, Plus and Ebsilon ($t_2 = $ const) and approximately by 5% in Ebsilon ($t_2 = $ var).

According to calculation of chemical energy rate of combustion Q_{cc} mentioned in Tables 2 and 9, this value in used software is approximately 307 kW and 336 kW for mixture 1 and methane, respectively. The results represent that the net efficiency of the system is 44%, 43.8%, 44.16%, and 43.07% for mixture 1 in Aspen Hysys, Plus, Ebsilon (including $t_2 = $ const and $t_2 = $ variable), respectively. These values are 43.32%, 43.05%, 43.12%, and 42.8% for methane for the mentioned software, respectively. It can be found from the results of Aspen Hysys, Plus, and Ebsilon ($t_2 = $ const) that, at the same assumption, changing fuels from methane to mixture results in rising the net efficiency of system from 1.5 to 2.4%.

The main source of the difference in the results obtained in Sections 3.1 and 3.2 is the fact that the specific heat was determined differently. This becomes apparent in the temperature results after pumps, compressors, and expanders.

Table 9. Effect of different fuels.

Parameter	Symbol	Unit	Mixture 1 (Syngas)	Methane
Temperature at the WCC outlet	$t_2 = $ var	°C	1073	1091
	$t_2 = $ const	°C	1100	1100
Fuel mass flow $(\dot{m}_{1-fuel})$	Aspen Hysys / Aspen Plus / Ebsilon $t_2 = $ var / Ebsilon $t_2 = $ const	g/s	18.00	6.72
Oxygen mass flow $(\dot{m}_{1-O_2})$	Aspen Hysys / Aspen Plus / Ebsilon $t_2 = $ var / Ebsilon $t_2 = $ const	g/s	23.2 / 22.4	26.8
Water mass flow $(\dot{m}_{1-H_2O})$	Aspen Hysys / Aspen Plus / Ebsilon $t_2 = $ var / Ebsilon $t_2 = $ const	g/s	58.8 / 59.6	66.48
Exhaust temperature after HE (t_5)	Aspen Hysys / Aspen Plus / Ebsilon $t_2 = $ var / Ebsilon $t_2 = $ const	°C	178.60 / 183.58 / 167.64 / 147.3	161.10 / 165.82 / 161.47 / 155.65
Turbine power GT (N_{GT})	Aspen Hysys / Aspen Plus / Ebsilon $t_2 = $ var / Ebsilon $t_2 = $ const	kW	88.73 / 89.30 / 87.67 / 89.53	92.93 / 93.20 / 92.65 / 93.26
Turbine power GT^{bap} $\left(N_{GT-bap}\right)$	Aspen Hysys / Aspen Plus / Ebsilon $t_2 = $ var / Ebsilon $t_2 = $ const	kW	65.64 / 64.9 / 63.69 / 65.20	68.49 / 67.52 / 67.11 / 67.63
Combined turbines gross power (N_t)	Aspen Hysys / Aspen Plus / Ebsilon $t_2 = $ var / Ebsilon $t_2 = $ const	kW	154.37 / 154.20 / 151.36 / 154.72	161.42 / 160.72 / 159.76 / 160.89
Power for own needs (N_{cp})	Aspen Hysys / Aspen Plus / Ebsilon $t_2 = $ var / Ebsilon $t_2 = $ const	kW	19.12 / 19.30 / 18.94	15.75 / 16 / 15.954 / 15.954
Chemical energy rate of combustion $\dot{Q}_{cc}$	Aspen Hysys / Aspen Plus / Ebsilon $t_2 = $ var / Ebsilon $t_2 = $ const	kW	307.42 / 307.45	336.23 / 336.1
Net efficiency (η_{net})	Aspen Hysys / Aspen Plus / Ebsilon $t_2 = $ var / Ebsilon $t_2 = $ const	%	44.00 / 43.88 / 43.07 / 44.16	43.32 / 43.05 / 42.8 / 43.12
Gross efficiency (η_g)	Aspen Hysys / Aspen Plus / Ebsilon $t_2 = $ var / Ebsilon $t_2 = $ const	%	50.21 / 50.16 / 49.23 / 50.32	48.01 / 47.81 / 47.55 / 47.86

3.3. N_2, NO, N_2O and NO_2 Formation and Influence on Temperature

This subsection is intended to indicate the effect of the ammonia combustion reaction on the temperature in the combustion chamber. Due to the fact that Ebsilon is mainly adapted to flow analyses with less flexibility in setting combustion data, this subsection is mainly based on results from Aspen Plus and Aspen Hysys. General chemical reactions in ammonia combustion are as follows:

$$4NH_3 + 3O_2 \rightarrow 2N_2 + 6H_2O \tag{11}$$

$$4NH_3 + 5O_2 \rightarrow 4NO + 6H_2O \tag{12}$$

$$4NH_3 + 7O_2 \rightarrow 4NO_2 + 6H_2O \tag{13}$$

$$2NH_3 + 2O_2 \rightarrow N_2O + 3H_2O \tag{14}$$

In Ebsilon, the basic reaction is the conversion of ammonia to nitrogen according to reaction (11). Aspen Plus, on the other hand, assigns the basic reaction to the conversion of ammonia to nitric oxide according to stoichiometric Equation (12), by default. However, due to the fact that different results are obtained in Tables 7 and 9, it was worthwhile to trace the other possibilities for the conversion of ammonia in the presence of oxygen and hence a set of (14) equations.

HYSYS calculates and displays the heat of reactions in the reaction heat cell. Table 10 depicts the reaction heat of different mentioned reactions. In this case, all of the reaction heat cells are negative, indicating that the reaction produces heat (exothermic). In thermodynamics, the term exothermic process describes a process or reaction that releases energy from the system to its surroundings, usually in the form of heat, but also in a form of light (e.g., a spark, flame, or flash), electricity (e.g., a battery), or sound (e.g., explosion heard when burning hydrogen). So, reactions 11 to 14 release 3.2×10^5 (kJ/kgmol), 2.3×10^5 (kJ/kgmol), 2.8×10^5 (kJ/kgmol), and 2.8×10^5 (kJ/kgmol), respectively. It can be understood that if all of these reactions could occur, reaction 11 releases the highest value of energy. This would take precedence in comparison to other reactions. Meanwhile, it is assumed that, in combined reaction, including reactions of N_2, N_2O, NO, and NO_2, each reaction is with ammonia conversion factor of 0.25.

Table 10. Reaction heat for different reactions obtained by Hysys.

Reactions	Heat of Reaction *, kJ/kgmol
$4NH_3 + 3O_2 \rightarrow 2N_2 + 6H_2O$	-3.2×10^5
$4NH_3 + 5O_2 \rightarrow 4NO + 6H_2O$	-2.3×10^5
$4NH_3 + 7O_2 \rightarrow 4NO_2 + 6H_2O$	-2.8×10^5
$2NH_3 + 2O_2 \rightarrow N_2O + 3H_2O$	-2.8×10^5

* at 25 °C.

The effect of an ammonia combustion reaction on the temperature is shown in Table 11. It shows the results obtained with Aspen Plus and Aspen Hysys. On the basis of the given chemical reactions and the obtained results of temperature and elemental compositions downstream of the combustion chamber, it should be concluded that the highest energy effect accompanies the formation of NO_2, followed by the formation of N_2 and N_2O, successively, and the lowest temperature is downstream of the combustion chamber after the formation of NO.

Table 11. 18 g/s mixture 1 (syngas) with NH_3 combustion to NO, NO_2, N_2, N_2O under stoichiometric conditions (100 g/s exhaust).

Parameter	Symbol	Unit	Combined *	N_2	N_2O	NO	NO_2
Temperature at the WCC outlet t_2 = var	Aspen Hysys		1107	1106	1104	1100	1116
	Aspen Plus	°C	1106	1105	1103	1100	1115
	Ebsilon		n.a.	1100	n.a.	n.a.	n.a.
Fuel mass flow $(\dot{m}_{1-fuel})$	Aspen Hysys Aspen Plus Ebsilon	g/s			18.00		

Table 11. *Cont.*

Parameter	Symbol	Unit	Combined *	N_2	N_2O	NO	NO_2
Oxygen mass flow $(\dot{m}_{1-O_2})$	Aspen Hysys	g/s	23.13	22.72	22.96	23.19	23.66
	Aspen Plus		23.13	22.72	22.96	23.19	23.66
	Ebsilon		n.a.	22.4	n.a.	n.a.	n.a.
Water mass flow $(\dot{m}_{1-H_2O})$	Aspen Hysys	g/s	58.86	59.28	59.04	58.80	58.33
	Aspen Plus		58.86	59.28	59.04	58.80	58.33
	Ebsilon		n.a.	59.6	n.a.	n.a.	n.a.
Exhaust temperature after HE (t_5)	Aspen Hysys	°C	187.9	186.1	186.3	183.60	195.00
	Aspen Plus		187.33	185.57	185.81	183.55	194.44
	Ebsilon		n.a.	147.3	n.a.	n.a.	n.a.
Turbine power GT (N_{GT})	Aspen Hysys	kW	89.40	89.61	89.26	89.05	89.66
	Aspen Plus		89.58	89.79	89.44	89.25	89.85
	Ebsilon		n.a.	89.53	n.a.	n.a.	n.a.
Turbine power GT^{bap} (N_{GT-bap})	Aspen Hysys	kW	65.46	65.58	65.37	65.14	65.75
	Aspen Plus		65.23	65.35	65.13	64.92	65.53
	Ebsilon		n.a.	65.2	n.a.	n.a.	n.a.
Combined turbines gross power (N_t)	Aspen Hysys	kW	154.9	155.2	154.6	154.2	155.4
	Aspen Plus		154.8	155.1	154.6	154.2	155.4
	Ebsilon		n.a.	154.72	n.a.	n.a.	n.a.
Power for own needs (N_{cp})	Aspen Hysys	kW	19.03	18.94	18.99	19.05	19.15
	Aspen Plus		19.28	19.19	19.25	19.30	19.40
	Ebsilon		n.a.	18.94	n.a.	n.a.	n.a.
Chemical energy rate of combustion (Q_{CC})	Aspen Hysys	kW			307.49		
	Aspen Plus						
	Ebsilon				307.45		
Net efficiency (η_{net})	Aspen Hysys	%	44.18	44.32	44.12	43.96	44.32
	Aspen Plus		44.08	44.21	44.01	43.86	44.22
	Ebsilon		n.a.	44.16	n.a.	n.a.	n.a.
Gross efficiency (η_g)	Aspen Hysys	%	50.37	50.48	50.30	50.16	50.55
	Aspen Plus		50.35	50.45	50.27	50.14	50.53
	Ebsilon		n.a.	50.32	n.a.	n.a.	n.a.
N_2 mass flow	Aspen Hysys	g/s	0.10	0.41			
	Aspen Plus		0.10	0.41	-	-	-
	Ebsilon		n.a.	0.31			
N_2O mass flow	Aspen Hysys	g/s	0.16		0.65		
	Aspen Plus		0.16	-	0.65	-	-
	Ebsilon		n.a.		n.a.		
NO mass flow	Aspen Hysys	g/s	0.22			0.89	
	Aspen Plus		0.22	-	-	0.89	-
	Ebsilon		n.a.			n.a.	
NO_2 mass flow	Aspen Hysys	g/s	0.34				1.36
	Aspen Plus		0.34	-	-	1.36
	Ebsilon		n.a.				n.a.

* Combined—each reaction with ammonia conversion factor of 0.25 (0.25 × 4 = 1).

4. PFD with Spray Ejector Condenser

In Figure 6, the extended version of the "PFD0" cycle is presented, shown in the previous chapter. The developed cycle "PFD1" includes additionally fuel preparation and carbon capture storage (CCS) units. Fuel comes out from the gasifier (R) as a product of a thermochemical process transformation of supplied dry sewage sludge in the presence of a gasifying agent. The gasifying agent is released after GT with optional release from a carbon capture unit (CCU) at an ambient pressure, consisting of a mixture of steam and

CO_2. The gasifying agent properties, such as content of CO_2, steam, and its temperature or pressure, can be controlled as required. An oxygen compressor (CO_2) is supplied from an air separation unit (ASU). A spray ejector condenser (SEC) sucks the exhaust from the heat exchanger 1 (HE1), while the motive fluid is supplied to SEC through the dedicated pump (PSEC).

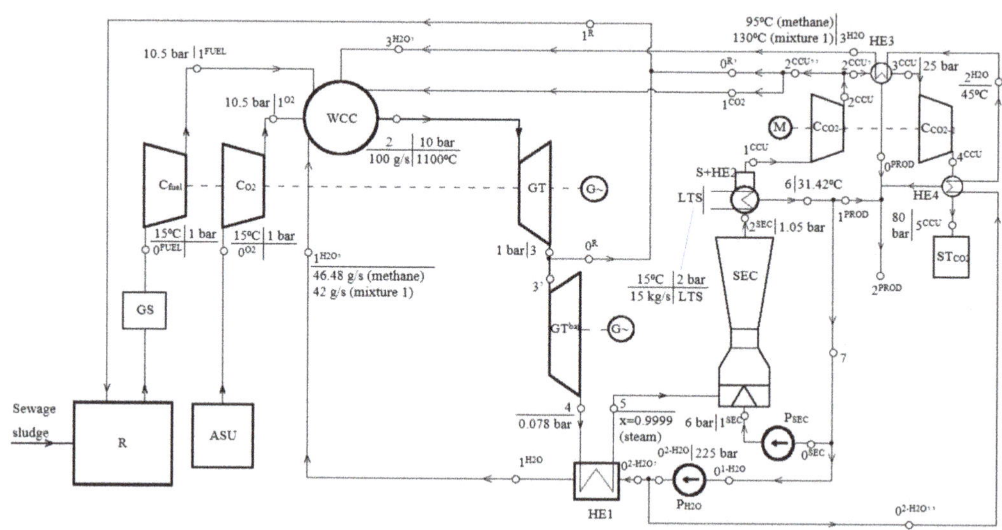

Figure 6. Process flow diagram of a gas mixture cycle—a steam-gas turbine system, where: WCC—wet combustion chamber; SEC—spray ejector condenser; R—gasifier (Reactor); GT—gas–steam turbine; GT^{bap}—gas–steam turbine—below ambient pressure; C_{fuel}—fuel compressor; C_{O_2}—oxygen compressor; $C_{CO_2-1,2}$—CO_2 capture unit compressors 1 and 2; P_{H_2O}—water pump supplying supercritical water; P_{SEC}—water pump supplying SEC, S + HE2—separator with heat exchanger 2; HE 1, 3, and 4—heat exchanger 1, 3, and 4; ASU—air separation unit; GS—gas scrubber; G~—power generators; M—motor; LTS—low-temperature source, ST_{CO_2}—CO_2 storage tank. Nodal points—general thermodynamic cycle: 0^{FUEL}, 0^{O_2}, 1^{FUEL}, 1^{O_2}, 2, 3, 3', 4, 5, 6, 7; optional: 2'; CO_2 capture unit: 1^{CCU}, 2^{CCU}, 3^{CCU}, 4^{CCU}, 5^{CCU}; optional: $2^{CCU'}$, $2^{CCU''}$; SEC: 0^{SEC}, 1^{SEC}, 2^{SEC}; gasifying agent supply: 0^R; optional: $0^{R'}$, 1^R; water production: 0^{PROD}, $1^{PROD'}$, 2^{PROD}; optional CO_2 injection to WCC: 1^{CO_2}; water supply: 0^{1-H_2O}, 0^{2-H_2O}, $0^{2-H_2O'}$, $0^{2-H_2O''}$, 1^{H_2O}, 2^{H_2O}, 3^{H_2O}; optional: $1^{H_2O'}$, $1^{H_2O''}$, $3^{H_2O'}$, $3^{H_2O''}$.

The outlet mixture of condensed steam and moist CO_2 vapor from SEC is directed to the separator with heat exchanger 2 (S + HE2), where low temperature source (LTS) is supplied and separation of CO_2 takes place. Water from HE2 is directed to PH2O and PSEC, while excess water is discharged out of the plant. Humid CO_2 vapor from the separator is directed to the CCU whereby, after each CO_2 compressor 1 and 2 (CCO2-1 and CCO2-2), there are intercoolers heat exchangers 3 and 4 (HE3 and HE4) with decantation which are supplied with water supplied from PH2O. Water after heating in CCU is directed to WCC where it reaches supercritical conditions. A partial release of CO_2 vapor can be used as a gasifying agent to the gasifier (R) or to WCC to manipulate and obtain the desired chemical reactions pathway. CO_2 vapor is directed to CO_2 storage tank ($STCO_2$) or can be used for other processes, such as methanol production.

The simulation models of "PFD1" developed in different computing codes are presented in Figures 7–9 (Aspen Hysys—Figure 7, Aspen Plus—Figure 8, Ebsilon—Figure 9), with most significant assumptions and calculated values in nodal points. Models do not contain the part connected with fuel preparation (gasifier and air separation unit). The main difference between models was approach to CCS part, especially with SEC modeling. The Ebsilon model (Figure 9) uses a spray ejector component, whereas other cycles define SEC operation through indirect models (direct-contact heat exchanger model, as shown in Figures 7 and 8). Moreover, cycles have a different arrangement of circulating water, which is extracted from exhaust gases. Next, through various configuration systems of heat exchanger, pumps are directed to WWC or SEC. One of the differences of simulation between Aspen Hysys and Plus is to consider decantation of heat exchangers. More accurately, separators are assumed in Aspen Hysys for decantation of heat exchangers, named decantation 1 and decantation 2, as seen in Figure 7.

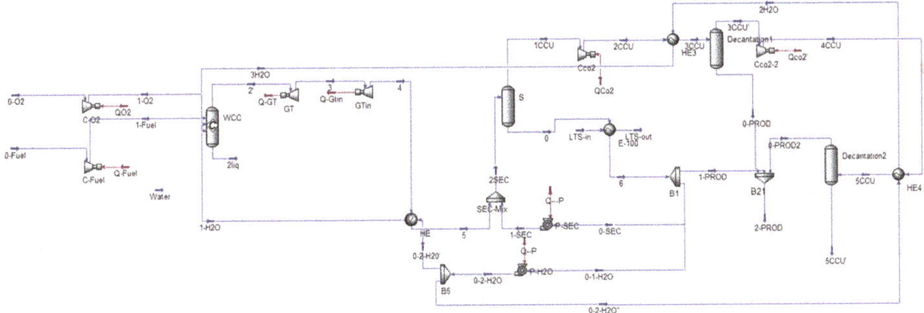

Figure 7. Simulation of PFD1 by Aspen HYSYS.

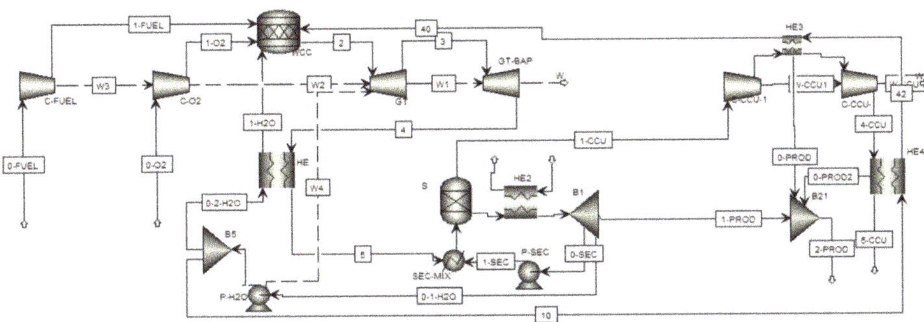

Figure 8. Simulation of PFD1 by Aspen Plus.

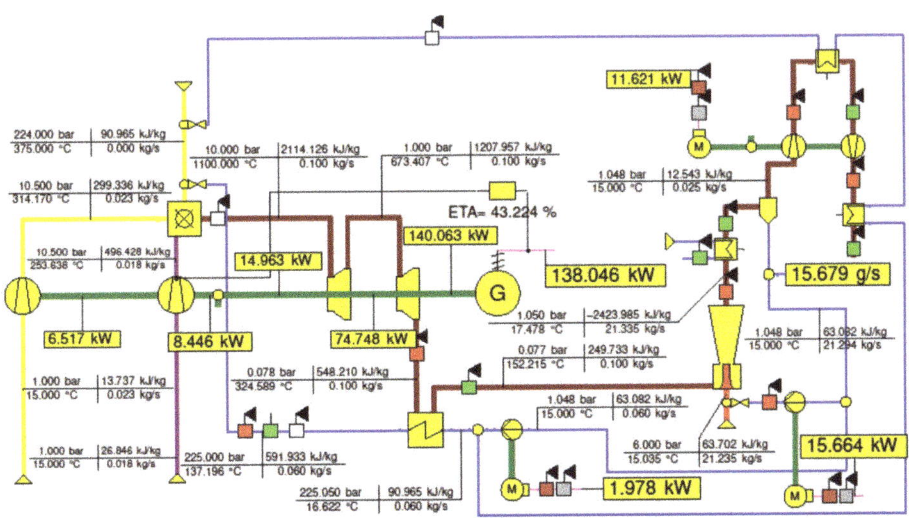

Figure 9. Simulation of PFD1 by by Ebsilon.

In addition, although five reactions are available for mixture 1 fuel as a default in Aspen Plus, there is a need to define these five reactions in Aspen Hysys.

Subsection

A power plant design based on "PFD1" presented in this paper supplied with mixture 1 syngas fuel is assumed to be the target operation configuration. Table 12 shows the results in the case of Aspen Plus and Aspen Hysys for 10 bar and 1100 °C in WCC, and a comparison with methane. However, for Ebsilon, it is presented with a temperature lower than 1100 °C due to the fact that a higher level of similarities in efficiency was obtained.

Table 12. Results for mixture 1 and methane—"PFD1"—summary table.

Parameter	Symbol	Unit	Value Aspen Plus		Value Aspen HYSYS		Value Ebsilon	
			Mixture 1	Methane	Mixture 1	Methane	Mixture 1	Methane
Fuel type	—	-	Mixture 1	Methane	Mixture 1	Methane	Mixture 1	Methane
Fuel mass flow	$\dot{m}_{1-fuel}$	g/s	16.68	6.23	16.68	6.23	16.68	6.23
Oxygen mass flow	$\dot{m}_{1-O_2}$	g/s	21.21	24.86	21.21	24.86	20.76	24.85
Water mass flow	$\dot{m}_{1-H_2O}$	g/s	62.11	68.91	62.11	68.91	62.56	68.92
CO_2 mass flow in exhaust	$\dot{m}_{2-CO_2}$	g/s	22.68	17.10	22.68	17.10	22.68	17.09
NO mass flow in exhaust	$\dot{m}_{2-NO}$	g/s	0.82	-	0.82	-	-	-
Water mass flow in exhaust	$\dot{m}_{2-H_2O}$	g/s	76.50	82.90	76.50	82.90	76.93	82.91
Water production	$\dot{m}_{p-H_2O}$	g/s	14.38	14.00	14.38	14.00	14.226	13.876
Exhaust temperature (before regenerative HE1, after GT^{bap})	t_4	°C	322.11	317.95	321.4	317.1	288.0	283.04
Exhaust temperature (after regenerative HE1, x = 0.9999)	t_5	°C	41.83	41.83	38.95	39.73	38.95	39.73
Turbine power GT	N_{GT}	kW	90.3	94.0	91.05	95.38	86.00	89.168
Turbine power GT^{bap}	N_{GT-bap}	kW	65.6	68.0	66.14	69.04	62.16	64.19
Combined turbines gross power	N_t	kW	155.9	162.0	157.19	164.42	148.16	153.358
Optimistic SEC Pump power consumption (x = 0 in mixing part of SEC)	$N_{P-SEC,o}$	kW	17.79	12.94	17.79	12.89	14.57	14.84
Not optimistic SEC Pump power consumption (x = 0.25 in mixing part of SEC)	$N_{P-SEC,n}$	kW	54.93	53.19	54.93	53.18		
Power for own needs with optimistic SEC	$N_{cp,o}$	kW	43.61	32.49	43.59	32.62	41.22	35.612
Power for own needs with optimistic SEC	$N_{cp,no}$	kW	80.75	72.74	80.63	72.91		
Chemical energy rate of combustion	$\dot{Q}_{CC}$	kW	284.86	311.82	284.88	311.72	284.97	311.59
Net efficiency with optimistic SEC	$\eta_{net,o}$	%	39.43	41.54	39.91	41.58	37.53	37.8
Net efficiency with not optimistic SEC	$\eta_{net,no}$	%	26.40	28.63	26.87	28.66		
Gross efficiency	η_g	%	54.74	51.96	55.18	52.10	52.00	49.22

Figures 10 and 11 show the graphs plotted for combined turbines power, chemical rate of combustion, CO_2 fraction, exhaust water fraction, water production, and gross efficiency for mixture 1 and methane obtained from various computing codes, such as Aspen Plus, Aspen Hysys, and Ebsilon. The maximum combined power generated by turbines is 164.42 kW from methane at the mass flow of water 14.0 g/s obtained from Aspen Hysys. The minimum combined power generated by turbines is 148.16 kW from mixture 1 at the mass flow of water 14.226 g/s produced due to combustion. This result was obtained from Ebsilon. In Figure 11, the maximum combined power generated by turbines is 164.42 kW from methane at CO_2 in the exhaust of 17.1% mass obtained from Aspen Hysys. The minimum combined power generated by turbines is 148.16 kW from mixture 1 at CO_2 in the exhaust of 22.68% mass obtained from calculation in Ebsilon. Combined turbine power output was much higher in case of methane combustion in all computing codes. However, the power output obtained by thermodynamic analyses in Ebsilon was obtained at a lower level because the temperature of the beginning of the expansion started from a lower level. The CO_2 and H_2O content change depends on fuel composition in different ways. Thus, the relationship between the power output obtained and the composition of the flue gases composition that flow through its successive stages becomes apparent.

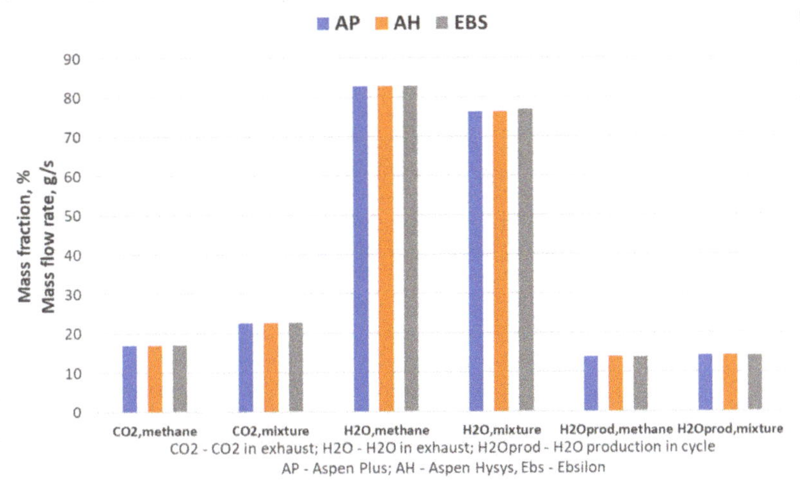

Figure 10. Exhaust CO_2, H_2O fraction and mass flow of water production in the cycle for various computing codes (AP—Aspen Plus, AH—Aspen Hysys, Ebs—Ebsilon).

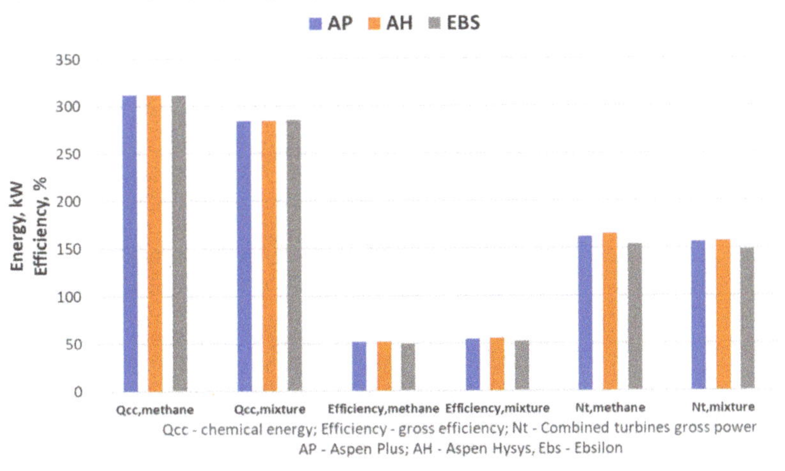

Figure 11. Combined turbine power, chemical energy of mixture 1 and methane combustion, and gross efficiency for various computing codes (AP—Aspen Plus, AH—Aspen Hysys, Ebs—Ebsilon).

The graphs in Figure 11 are plotted for chemical rate of combustion, combined power, and gross efficiency for mixture 1 and methane from Aspen Plus, Aspen Hysys, and Ebsilon, respectively. It should be noted that the results obtained from Aspen Plus and Aspen Hysys indicate the same values. The chemical energy rate of combustion was similar in the case of Ebsilon software.

The highest efficiency was calculated for mixture 1 in Apen Plus and Aspen Hysys (similar values, respectively: 54.74% and 55.18%). Lower efficiencies were achieved in the case of the Ebsilon computing code (52.00% for mixture 1 and 49.22% for methane).

For gross efficiency, there is a similar relationship for turbine power output, but additionally the chemical energy rate of the fuel is taken into account, which ultimately results in higher efficiencies for flue gases with increased steam production. A similar trend is observed for increased CO_2 in the flue gas. Thus, in order to clearly determine the effect of the fuel mixture on the performance of turbine and the entire nCO2PP cycle, a wider range of fuels would have to be studied—but this was not the purpose of the paper. First of all, it should be stated that there is a slight influence of the software used on the results obtained, but the basic tendencies are the same, which makes it possible to analyze various types of thermodynamic cycles.

The values of fuel mass flow, oxygen mass flow, and water mass flow of mixture 1 (syngas) and methane used in Aspen Plus and Aspen HYSYS are the same, but are a not so different from the values used in Ebsilon. So, this impacts the simulation of process flow diagram 1 ("PFD1") and the values obtained in Ebsilon is comparatively different from the values obtained from Aspen Plus and Aspen HYSYS. There are several reasons why the results may not be exactly the same. Firstly, there are some differences regarding simulation models, and procedures adopted inside the model preparation.

In calculations using the model developed thanks to the Ebsilon software (Figure 9), the first assumption is the mass flow rate of fuel together with an assumption of stoichiometric combustion inside WCC. Next, the amount of oxygen is calculated. The mass flow rate exhaust gases depends on the amount of cooling water to combustion chamber which is equal to 100 g/s. Based on this procedure and assumptions, nodal values in the thermal cycle can be computed.

Another difference in simulation between Ebsilon, Aspen Hysys, and Plus is SEC. An operating principle of the spray ejector condenser (SEC) shown in Figure 12 is described as follows.

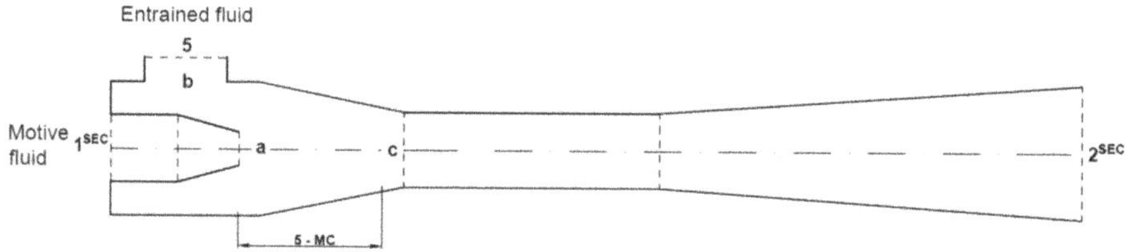

Figure 12. Spray ejector condenser 1^{SEC}: Nozzle inlet, a: Nozzle outlet, a–c: Mixing chamber, b: Suction chamber, c: Throat (part of Diffuser), c-2^{SEC}: Diffuser, 5 Inlet of entrained fluid, 5-MC: Entrained fluid in mixing chamber.

Motive fluid in subsonic flow enters the nozzle (1^{SEC}) in Figure 12, which has a decreasing cross section area in which motive fluid is accelerated, while pressure energy is converted to velocity energy. Sonic flow velocity is reached at the same time when a minimum area of the nozzle (1^{SEC}) is reached. In supersonic flow, the nozzle (1^{SEC}) is an increasing area device. Entrained fluid (5) enters the suction chamber (b) and increases its velocity. The motive fluid and entrained fluid (5-MC) mix together in a mixing chamber (a-c). The mixture is directed to diffuser—throat (c) in supersonic flow in a decreasing area. The diffuser (c-2^{SEC}) in subsonic flow has an increasing area and converts velocity energy to pressure energy. The proper design of SEC is important for the feasibility in operation for a particular case of required conditions. Basically, ejectors are designed using a lot of empirical correlations and any information related to their design is not available in the public domain. It would be recommended to rely on the design characteristics.

In order to improve the design, efficiency, and feasibility of SEC, similar characteristics could be obtained using CFD calculations and data from laboratory experiments for the purposes of the paper. Elongation of the nozzle part can possibly contribute to an increase in efficiency. Because of high compression ratio, further research has to be conducted to decide whether a single configuration or several ejectors in multiple stages, and vertical or horizontal alignment of an ejector, would be preferred.

In this study, the motive fluid has to be H_2O (1^{SEC}) while the entrained fluid (5) is the mixture of CO_2 and H_2O. Both CO_2 and H_2O occupy a large volume, causing a decrease in efficiency. For the optimum case with novel approach steam, H_2O would immediately be condensed (5-MC) in mixing chamber of SEC, contributing to an increase in efficiency at the same time, but in the less favorable case steam would be partially condensed in SEC, resulting in a decrease in efficiency due to the increase in a required motive fluid mass flow.

Although the ejector was available in Ebsilon for a simulation of SEC, the mixer was used in Aspen Hysys and Plus due to lack of ejector in mentioned software. So, the desired results of water when used in mixer are obtained according to Equation (15).

$$\chi = \frac{\dot{V}_{5-MC}}{\dot{V}_{1_{SEC}}} , - \tag{15}$$

where χ is volumetric entrainment ratio considered. V_{5-MC} demonstrates suction gas–fluid volume flow to the mixing chamber of SEC (m^3/s) and $V_1{}^{SEC}$ is the motive fluid volume flow (m^3/s).

Moreover, the assumptions concerning fuels were a little different (low heating value, fuel inlet composition). This was because computing codes used various physical tables regarding fuel properties. The iterative method of calculating the problem also seems to be a crucial factor.

5. Negative Emission Power Plant Effect

Currently, sewage sludge is considered as a biomass, according to the new Polish Act on Renewable Energy Sources of 20 February 2015 and its novel version of 19 July 2019. The possibility to utilize sewage sludge in gasification process is an additional advantage of the proposed solution. A comparison of the emissivity of the systems for the different options presented in this report is summarized in Table 13. Of the parameters listed in the table, two which determine carbon dioxide emissions are especially noteworthy, namely:

$$eCO_2 = \frac{\dot{m}_{2-CO_2}}{N_t - N_{cp}} 3600 \tag{16}$$

$$\eta_{net} \cdot eCO_2 = \frac{N_t - N_{cp}}{\dot{Q}_{CC}} \frac{\dot{m}_{2-CO_2}}{N_t - N_{cp}} 3600 = \frac{\dot{m}_{2-CO_2}}{\dot{Q}_{CC}} 3600 \tag{17}$$

Negative emissions of CO_2 were counted based on two parameters defined in Equations (15) and (16). Firstly, in (15), the specific CO_2 emission is given, which is the quotient of the CO_2 capture mass flow rate with respect to the net power. The net power was classically defined as the difference in the turbine-generated power N_t and the demand power N_{cp}. This definition is also found in the works of authors, such as [57] or [58], in relation to cycles with CO_2 capture. Secondly, Equation (16) defines the product of the efficiency of the whole cycle and the specific CO_2 emitted. Additionally, after simplification, this parameter directly expresses the relative emissivity related to the chemical energy rate. In a traditional view, both parameters (Equations (15) and (16)) show the emissions of the unit, but in the case of nCO2PP they are an indicator of the negative emissions related to the electrical energy obtained from the cycle or to the chemical energy supplied to the cycle, respectively. The results in Table 13 were selected for Aspen Plus and Hysys as the least optimistic of the previous results in Section 4.

Table 13. Negative emission power plant effect—results for Aspen Plus.

Parameter	Software	Symbol	Unit	Methane PP -Conventional	Methane PFD with SEC Zero-Emission	Mixture PFD with SEC nCO_2PP
Net efficiency with optimistic SEC	Aspen Plus	η_{net}	%	47.1	41.5	39.4
	Aspen Hysys			47.1	41.5	39.9
CO2 mass flow in exhaust	Aspen Plus	$\dot{m}_{2-CO_2}$	g/s	17.1	17.1	22.7
	Aspen Hysys			17.1	17.1	22.7
Power for own needs with optimistic SEC	Aspen Plus	N_{cp}	kW	15.0	32.5	43.6
	Aspen Hysys			15.0	32.6	43.6
Turbine power output	Aspen Plus	N_t	kW	162.0	162.0	155.9
	Aspen Hysys			162.0	164	157.1
Chemical energy rate of combustion	Aspen Plus	$\dot{Q}_{CC}$	kW	311.8	311.8	284.9
	Aspen Hysys			311.8	311.8	284.9
Emission of carbon dioxide	Aspen Plus	eCO_2	kg/MWh	418.78	0.0	−727.12
	Aspen Hysys			418.78	0.0	−720.0
Relative emissivity of carbon dioxide	Aspen Plus	$\eta_{net} \cdot eCO_2$	%kg/MWh	197.42	0.0	−286.70
	Aspen Hysys			197.42	0.0	−286.70
Avoided emission of carbon dioxide	Aspen Plus	Avoid eCO_2	kg/MWh	0.00	475.33	1454.23
	Aspen Hysys			0.00	476.22	1440
Avoided relative emissivity of carbon dioxide	Aspen Plus	Avoid $\eta_{net} \cdot eCO_2$	%kg/MWh	0.00	197.45	573.40
	Aspen Hysys			0.00	197.54	573.40
Specific Primary Energy Consumption for Carbone Avoided	Aspen Plus	$SPECCA$	MJ/kgCO2	NA	0.999	0.822
	Aspen Hysys			NA	0.999	0.822

As shown in Table 13 in the conventional cycle where methane is burnt, the emissivity related to the electrical energy eCO_2 for both Aspen Plus and Hysys is 418.78 kg/MWh$_{el}$ and, in case of emissivity related to the chemical energy, $\eta_{net} \cdot eCO_2$ is 197.42 kg/MW$_{ch}$. An additional set of equipment should be used to avoid carbon dioxide emissions. The emissions of CO_2, relative to the power output, for the combustion of methane in the novel power plant, outlined in this paper (Table 13), were slightly lower in comparison to the reference case used by Saari et al. [22] (482 kg/MWh$_{el}$). However, in terms of negative emissions achieved with producer gas from gasification of sewage sludge, the novel power plant concept significantly outperformed chemical looping with oxygen uncoupling (CLOU) plant, as proposed by Saari et al. [22] (13 kg/MWh$_{el}$). In the zero-carbon unit, on the other hand, we capture carbon dioxide and, thus, avoid emissivity related to the electrical energy at the level eCO_2 475.33 kg/MWh$_{el}$ in Aspen Plus and 476.22 in Aspen Hysys and, in the case of emissivity related to the chemical energy, we avoid $\eta_{net} \cdot eCO_2$ 197.45 kg/MWch and 197.54 kg/MW in Aspen Plus and Hysys, respectively. In the case of nCO2PP, the indicated coefficients are much more favorable. Both parameters show that the avoided emissivity of the block after carbon dioxide capture is equal to twice the absolute value of the previously determined numbers. Consequently, the avoided emissivity value for nCO2PP is about three times higher than that for zero-emission units.

The specific power consumption associates to the modelled oxygen generating station (ASU) is β = 0.248 kWh/kgO$_2$ (for comparison, the value of the energy intensity in a study by Gou et al. [59] is β = 0.247 kWh/kgO$_2$, while, in a study by Liu et al. [60], it is β = 0.250kWh/kgO$_2$).

To compare plants which include different capture efficiencies, regeneration temperature, and electrical efficiencies penalties, the specific primary energy consumption for carbone avoided is introduced according to other works [57,58]. For research, MEA is classified at the level 4.16; however, Bonalumi et al. improved this parameter to value 2.86 and 2.58 for chilled (with salts) and cooled (without salts), respectively. In the presented case (Table 13), the SPECCA value reaches 0.999 and 0.822 for zero-emmision and negative emission power plant, respectively.

6. Effect of Specific Heat Capacity

The specific heat capacity (also simply specific heat) of a substance is the heat capacity per unit mass of that substance. Here, we shall discuss the specific heat capacity using SI units (kJ/kg·K). Heat capacity can be expressed at a constant volume (c_v) or constant pressure (c_p). Specific heat capacity of mixture 1 at a constant pressure and volume as a function of temperature from 1 °C to 1300 °C, calculated using Aspen Hysys, Aspen Plus, and Ebsilon, are presented in Figures 13 and 14, respectively.

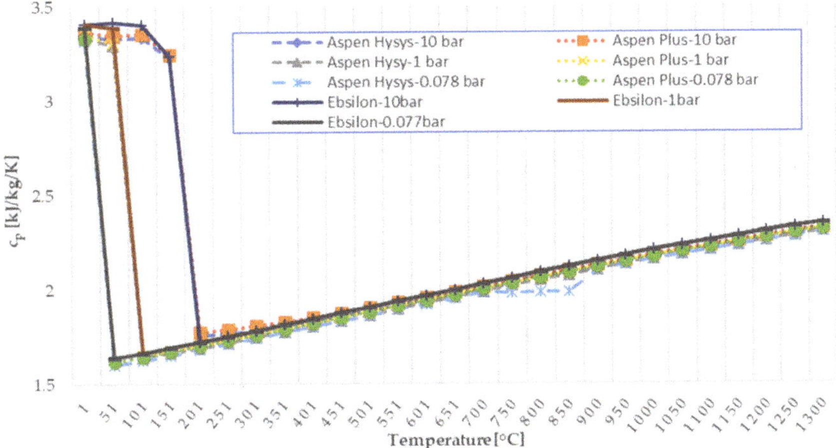

Figure 13. Distribution of heat capacity at constant pressure per temperature after WCC at different pressure in Aspen Hysys, Aspen Plus, and Ebsilon.

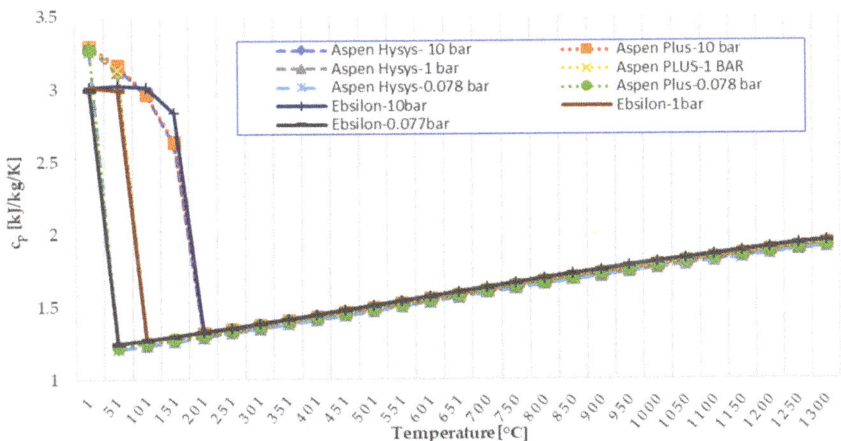

Figure 14. Distribution of heat capacity at constant volume per temperature after WCC at different pressure in Aspen Hysys, Aspen Plus and Ebsilon.

It can be regarded that c_p and c_v are faced with a decreased trend from 1 °C to 200 °C, so that the minimum value of c_p and c_v is 1.78 kJ/kg·K and 1.32 kJ/kg·K at 200 °C, respectively. After that, they increase with rising the temperature. In addition, for $T \geq 200$ °C, the specific heat capacity at constant volume (c_v) and constant pressure (c_p) remain approximately constant with increasing the pressure from 0.078 to 10 bar. Specific heat capacity (c_p) in nodal points for at 10 bar and 1100 °C for Aspen Hysys, Aspen Plus, and Ebsilon is represented. It may be viewed that the minimum value (0.92) of c_p belongs to oxygen, whilst the maximum one belongs to water (Figure 15). In addition, there is a direct relation between increasing temperature and c_p. The specific heat capacity values extracted from the codes are close to each other, but when the processes in the individual devices are taken into account, they affect the efficiency values of the whole cycle, in both considered versions of the nCO2PP cycle.

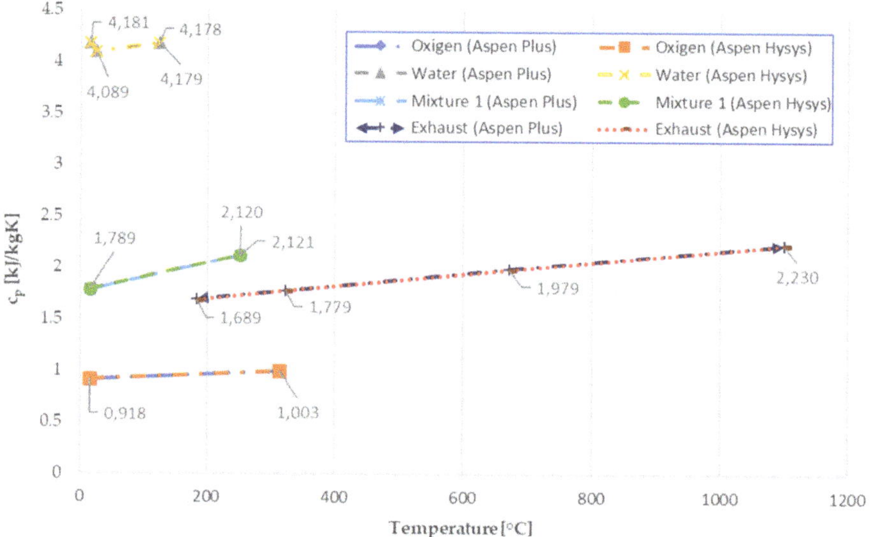

Figure 15. Specific heat capacity (c_p) in nodal points PFD0, mixture 1, WCC exhaust at 10 bar and 1100 °C for Aspen Hysys, Aspen Plus and Ebsilon.

7. Conclusions

The developed version of the cycle, called Process Flow Diagram 0 ("PFD0"), offered the possibility to perform a preliminary assessment of the main cycle parameters, generated power output, as well as temperature in the combustion chamber and at the turbine outlet. It could be concluded that the proposed design of the negative power plant could be considered feasible and competitive with other types BECCS plant, presented in the literature, especially when achievable negative emissions (-720 kg$_{CO_2}$/MWh$_{el}$) are taken into the account.

On the basis of the obtained results, the following key conclusions can be presented:

(1) The presented cycle "PFD0" allows generating approx. 150 kW for mixture 1 and 160 kW for methane in three considered software (Table 9).

(2) When inflicting the same mass flow rates (oxygen, water, mixture 1, or methane) and temperatures as in Ebsilon at the inlet to the combustion chamber, we obtain a temperature higher by 27 or 9 degrees Celsius or more in Aspen Plus and Aspen Hysys, and therefore the temperature at the exit from the WCC is 1073 or 1091 °C.

(3) On the other hand, when given the same mass flow rates (oxygen, water, mixture 1 or methane) and different temperatures downstream of the heat exchanger in the

Ebsilon, the temperature downstream of the combustion chamber can be constant, so the WCC plot is 1100 °C.

(4)   The trend is similar for mixture 1 and methane, but the differences are greater as we do not have the same set of reactions concerning the combustion chamber. In this case, the conversion of the ammonia combustion reaction to NO and H_2O to combustion to N_2 and H_2O gives a gain of 6 degrees Celsius more (see Table 11). In mixture 1, we have significant ammonia content why explains the large difference with respect to combustion in traditional chambers, where this influence is negligible.

(5)   An argument that a likely reason for the differences in the two codes are the different definitions, e.g., in one specific heat capacity of steam stabilized in Ebsilon and in Aspen specific heat capacity of steam following the P-R equation, is the fact that we obtain different temperatures after the pump and after the compressors with assumed isentropic efficiencies at the same level, at the same inlet temperatures, and at the same pressure rise (see Section 6).

Additionally, the proposed version of the cycle, called the process flow diagram 1 ("PFD1"), offered the possibility to perform a preliminary assessment of the main cycle parameters, consumed and generated powers, efficiencies, and temperatures in nodal points. The following conclusions can be drawn from Sections 4 and 5:

(1)   SEC significantly affects the efficiency of the cycle but provides the opportunity for carbon dioxide separation in the nCO2PP system.

(2)   Differences in the Aspen Plus, Aspen Hysys, and Ebsilon codes follow a similar trend.

(3)   In subsequent calculations, the modeling of the injector should be approached more extensively. For example, there should be more reliance on measurement results obtained from one's own experiment.

(4)   The possibility of a negative CO_2 emission power plant and the positive environmental impact of the proposed solution were demonstrated.

Author Contributions: Conceptualization, P.Z. and P.M.; methodology, P.Z. and P.M.; software, P.Z., P.M., M.A., K.S. and N.S.; validation, P.Z., P.M., H.P.-K., J.B. and D.M.; formal analysis, P.Z. and P.M.; investigation, P.Z., P.M., M.A., K.S. and N.S.; resources, P.Z. and P.M.; data curation, P.Z., P.M., M.A. and K.S.; writing—original draft preparation, P.Z., P.M., M.A., T.K., K.S. and N.S.; writing—review and editing, P.Z., P.M., H.P.-K., J.B., Ł.N. and D.M.; visualization, M.A., T.K., K.S.; supervision, H.P.-K., J.B., D.M.; project administration, D.M., P.Z., H.P-K. and P.M.; funding acquisition, D.M., P.Z. and P.M. All authors have read and agreed to the published version of the manuscript.

Funding: The research leading to these results has received funding from the Norway Grants 2014–2021 via the National Centre for Research and Development. Article has been prepared within the frame of the project: "Negative CO_2 emission gas power plant"—NOR/POLNORCCS/NEGATIVE-CO2-PP/0009/2019-00 which is co-financed by programme "Applied research" under the Norwegian Financial Mechanisms 2014–2021 POLNOR CCS 2019—Development of CO_2 capture solutions integrated in power and industry processes.

Conflicts of Interest: The authors declare no conflict of interest. The funders had no role in the design of the study; in the collection, analyses, or interpretation of data; in the writing of the manuscript, or in the decision to publish the results.

References

1.   United Nations Framework Convention on Climate Change—Paris Agreement. Available online: https://unfccc.int/process-and-meetings/the-paris-agreement/the-paris-agreement (accessed on 1 July 2021).

2.   Masson-Delmotte, V.; Zhai, P.; Pörtner, H.-O.; Roberts, D.; Skea, J.; Shukla, P.R.; Pirani, A.; Moufouma-Okia, W.; Péan, C.; Pidcock, R.; et al. *Global Warming of 1.5 °C*; IPCC—The Intergovernmental Panel on Climate Change: Geneva, Switzerland, 2019.

3.   Friedlingstein, P.; Jones, M.W.; O'Sullivan, M.; Andrew, R.M.; Hauck, J.; Peters, G.P.; Peters, W.; Pongratz, J.; Sitch, S.; Le Quéré, C.; et al. Global Carbon Budget 2019. *Earth Syst. Sci. Data* **2019**, *11*, 1783–1838. [CrossRef]

4.   Romanak, K.; Fridahl, M.; Dixon, T. Attitudes on Carbon Capture and Storage (CCS) as a Mitigation Technology within the UNFCCC. *Energies* **2021**, *14*, 629. [CrossRef]

5.   Hiremath, M.; Viebahn, P.; Samadi, S. An Integrated Comparative Assessment of Coal-Based Carbon Capture and Storage (CCS) Vis-à-Vis Renewable Energies in India's Low Carbon Electricity Transition Scenarios. *Energies* **2021**, *14*, 262. [CrossRef]

6. Sifat, N.S.; Haseli, Y. A Critical Review of CO_2 Capture Technologies and Prospects for Clean Power Generation. *Energies* **2019**, *12*, 4143. [CrossRef]
7. Qvist, S.; Gładysz, P.; Bartela, Ł.; Sowiżdżał, A. Retrofit Decarbonization of Coal Power Plants—A Case Study for Poland. *Energies* **2020**, *14*, 120. [CrossRef]
8. Gładysz, P.; Sowiżdżał, A.; Miecznik, M.; Hacaga, M.; Pająk, L. Techno-Economic Assessment of a Combined Heat and Power Plant Integrated with Carbon Dioxide Removal Technology: A Case Study for Central Poland. *Energies* **2020**, *13*, 2841. [CrossRef]
9. Gładysz, P.; Stanek, W.; Czarnowska, L.; Węcel, G.; Langørgen, Ø. Thermodynamic assessment of an integrated MILD oxyfuel combustion power plant. *Energy* **2017**, *137*, 761–774. [CrossRef]
10. Gładysz, P.; Stanek, W.; Czarnowska, L.; Sładek, S.; Szlęk, A. Thermo-ecological evaluation of an integrated MILD oxy-fuel combustion power plant with CO_2 capture, utilisation, and storage—A case study in Poland. *Energy* **2018**, *144*, 379–392. [CrossRef]
11. Míguez, J.L.; Porteiro, J.; Pérez-Orozco, R.; Gómez, M. Technology Evolution in Membrane-Based CCS. *Energies* **2018**, *11*, 3153. [CrossRef]
12. Sieradzka, M.; Gao, N.; Quan, C.; Mlonka-Mędrala, A.; Magdziarz, A. Biomass Thermochemical Conversion via Pyrolysis with Integrated CO_2 Capture. *Energies* **2020**, *13*, 1050. [CrossRef]
13. Cannone, S.F.; Lanzini, A.; Santarelli, M. A Review on CO_2 Capture Technologies with Focus on CO_2-Enhanced Methane Recovery from Hydrates. *Energies* **2021**, *14*, 387. [CrossRef]
14. Arora, A.; Kumar, A.; Bhattacharjee, G.; Kumar, P.; Balomajumder, C. Effect of different fixed bed media on the performance of sodium dodecyl sulfate for hydrate based CO_2 capture. *Mater. Des.* **2016**, *90*, 1186–1191. [CrossRef]
15. Arora, A.; Kumar, A.; Bhattacharjee, G.; Balomajumder, C.; Kumar, P. Hydrate-Based Carbon Capture Process: Assessment of Various Packed Bed Systems for Boosted Kinetics of Hydrate Formation. *J. Energy Resour. Technol.* **2020**, *143*, 033005. [CrossRef]
16. Detz, R.J.; van der Zwaan, B. Transitioning towards negative CO_2 emissions. *Energy Policy* **2019**, *133*, 110938. [CrossRef]
17. Lisbona, P.; Pascual, S.; Pérez, V. Evaluation of Synergies of a Biomass Power Plant and a Biogas Station with a Carbon Capture System. *Energies* **2021**, *14*, 908. [CrossRef]
18. Mendiara, T.; García-Labiano, F.; Abad, A.; Gayán, P.; de Diego, L.F.; Izquierdo, M.; Adánez, J. Negative CO_2 emissions through the use of biofuels in chemical looping technology: A review. *Appl. Energy* **2018**, *232*, 657–684. [CrossRef]
19. Bhui, B.; Vairakannu, P. Prospects and issues of integration of co-combustion of solid fuels (coal and biomass) in chemical looping technology. *J. Environ. Manag.* **2018**, *231*, 1241–1256. [CrossRef]
20. Lyngfelt, A.; Johansson, D.J.; Lindeberg, E. Negative CO_2 emissions—An analysis of the retention times required with respect to possible carbon leakage. *Int. J. Greenh. Gas Control* **2019**, *87*, 27–33. [CrossRef]
21. Niu, X.; Shen, L.; Jiang, S.; Gu, H.; Xiao, J. Combustion performance of sewage sludge in chemical looping combustion with bimetallic Cu–Fe oxygen carrier. *Chem. Eng. J.* **2016**, *294*, 185–192. [CrossRef]
22. Saari, J.; Peltola, P.; Tynjälä, T.; Hyppänen, T.; Kaikko, J.; Vakkilainen, E. High-Efficiency Bioenergy Carbon Capture Integrating Chemical Looping Combustion with Oxygen Uncoupling and a Large Cogeneration Plant. *Energies* **2020**, *13*, 3075. [CrossRef]
23. Buscheck, T.A.; Upadhye, R.S. Hybrid-energy approach enabled by heat storage and oxy-combustion to generate electricity with near-zero or negative CO_2 emissions. *Energy Convers. Manag.* **2021**, *244*, 114496. [CrossRef]
24. Pawlak-Kruczek, H.; Niedzwiecki, L.; Ostrycharczyk, M.; Czerep, M.; Plutecki, Z. Potential and methods for increasing the flexibility and efficiency of the lignite fired power unit, using integrated lignite drying. *Energy* **2019**, *181*, 1142–1151. [CrossRef]
25. Madejski, P.; Żymełka, P. Calculation methods of steam boiler operation factors under varying operating conditions with the use of computational thermodynamic modeling. *Energy* **2020**, *197*, 117221. [CrossRef]
26. Modliński, N.; Szczepanek, K.; Nabagło, D.; Madejski, P.; Modliński, Z. Mathematical procedure for predicting tube metal temperature in the second stage reheater of the operating flexibly steam boiler. *Appl. Therm. Eng.* **2019**, *146*, 854–865. [CrossRef]
27. Mączka, T.; Pawlak-Kruczek, H.; Niedzwiecki, L.; Ziaja, E.; Chorążyczewski, A. Plasma Assisted Combustion as a Cost-Effective Way for Balancing of Intermittent Sources: Techno-Economic Assessment for 200 MW_{el} Power Unit. *Energies* **2020**, *13*, 5056. [CrossRef]
28. Benato, A.; Bracco, S.; Stoppato, A.; Mirandola, A. LTE: A procedure to predict power plants dynamic behaviour and components lifetime reduction during transient operation. *Appl. Energy* **2016**, *162*, 880–891. [CrossRef]
29. Capron, M.; Stewart, J.; N'Yeurt, A.D.R.; Chambers, M.; Kim, J.; Yarish, C.; Jones, A.; Blaylock, R.; James, S.; Fuhrman, R.; et al. Restoring Pre-Industrial CO_2 Levels While Achieving Sustainable Development Goals. *Energies* **2020**, *13*, 4972. [CrossRef]
30. Cheng, F.; Small, A.A.; Colosi, L.M. The levelized cost of negative CO_2 emissions from thermochemical conversion of biomass coupled with carbon capture and storage. *Energy Convers. Manag.* **2021**, *237*, 114115. [CrossRef]
31. EU Carbon Price Hits Record 50 Euros per Tonne on Route to Climate Target | Reuters. Available online: https://www.reuters.com/business/energy/eu-carbon-price-tops-50-euros-first-time-2021-05-04/ (accessed on 7 August 2021).
32. Restrepo-Valencia, S.; Walter, A. Techno-Economic Assessment of Bio-Energy with Carbon Capture and Storage Systems in a Typical Sugarcane Mill in Brazil. *Energies* **2019**, *12*, 1129. [CrossRef]
33. Mikielewicz, D.; Wajs, J.; Ziółkowski, P.; Mikielewicz, J. Utilisation of waste heat from the power plant by use of the ORC aided with bleed steam and extra source of heat. *Energy* **2016**, *97*, 11–19. [CrossRef]
34. Szablowski, L.; Krawczyk, P.; Badyda, K.; Karellas, S.; Kakaras, E.; Bujalski, W. Energy and exergy analysis of adiabatic compressed air energy storage system. *Energy* **2017**, *138*, 12–18. [CrossRef]

35. Bartela, L.; Skorek-Osikowska, A.; Kotowicz, J. Economic analysis of a supercritical coal-fired CHP plant integrated with an absorption carbon capture installation. *Energy* **2014**, *64*, 513–523. [CrossRef]

36. Zymelka, P.; Szega, M.; Madejski, P. Techno-Economic Optimization of Electricity and Heat Production in a Gas-Fired Combined Heat and Power Plant with a Heat Accumulator. *J. Energy Resour. Technol.* **2020**, *142*, 022101. [CrossRef]

37. Kotowicz, J.; Job, M.; Brzęczek, M. The characteristics of ultramodern combined cycle power plants. *Energy* **2015**, *92*, 197–211. [CrossRef]

38. Topolski, J.; Badur, J. Efficiency of HRSG within a Combined Cycle with gasification and sequential combustion at GT26 Turbine. In Proceedings of the Second International Scientific Symposium Compower, Gdańsk, Poland, 4–7 September 2000; pp. 291–298.

39. Ziółkowski, P.; Badur, J.; Ziółkowski, P. An energetic analysis of a gas turbine with regenerative heating using turbine extraction at intermediate pressure - Brayton cycle advanced according to Szewalski's idea. *Energy* **2019**, *185*, 763–786. [CrossRef]

40. Głuch, J. Selected problems of determining an efficient operation standard in contemporary heat-and-flow diagnostics. *Pol. Marit. Res.* **2009**, *16*, 22–26. [CrossRef]

41. Ong'Iro, A.; Ugursal, V.; Al Taweel, A.; Lajeunesse, G. Thermodynamic simulation and evaluation of a steam CHP plant using ASPEN Plus. *Appl. Therm. Eng.* **1996**, *16*, 263–271. [CrossRef]

42. Liu, B.; Yang, X.-M.; Song, W.-L.; Lin, W.-G. Process simulation of formation and emission of NO and N2O during coal decoupling combustion in a circulating fluidized bed combustor using Aspen Plus. *Chem. Eng. Sci.* **2012**, *71*, 375–391. [CrossRef]

43. Jang, D.-H.; Kim, H.-T.; Lee, C.; Kim, S.-H. Kinetic analysis of catalytic coal gasification process in fixed bed condition using Aspen Plus. *Int. J. Hydrogen Energy* **2013**, *38*, 6021–6026. [CrossRef]

44. Nikoo, M.B.; Mahinpey, N. Simulation of biomass gasification in fluidized bed reactor using ASPEN PLUS. *Biomass Bioenergy* **2008**, *32*, 1245–1254. [CrossRef]

45. Damartzis, T.; Michailos, S.; Zabaniotou, A. Energetic assessment of a combined heat and power integrated biomass gasification–internal combustion engine system by using Aspen Plus®. *Fuel Process. Technol.* **2012**, *95*, 37–44. [CrossRef]

46. Steag Energy Services Ebsilon®Professional 15.00. Available online: https://www.ebsilon.com/ (accessed on 22 September 2021).

47. Madejski, P.; Żymełka, P. *Introduction to Computer Calculations and Simulation of Energy Systems Operation in STEAG Ebsilon®Professional*; Wydawnictwa AGH: Kraków, Poland, 2020.

48. Soares, J.; Oliveira, A.; Valenzuela, L. A dynamic model for once-through direct steam generation in linear focus solar collectors. *Renew. Energy* **2021**, *163*, 246–261. [CrossRef]

49. Yue, M.; Ma, G.; Shi, Y. Analysis of Gas Recirculation Influencing Factors of a Double Reheat 1000 MW Unit with the Reheat Steam Temperature under Control. *Energies* **2020**, *13*, 4253. [CrossRef]

50. Dahash, A.; Mieck, S.; Ochs, F.; Krautz, H.J. A comparative study of two simulation tools for the technical feasibility in terms of modeling district heating systems: An optimization case study. *Simul. Model. Pract. Theory* **2018**, *91*, 48–68. [CrossRef]

51. Mondal, S.K.; Uddin, M.F.; Majumder, S.; Pokhrel, J. HYSYS Simulation of Chemical Process Equipments. Available online: https://www.researchgate.net/publication/281608946_HYSYS_Simulation_of_Chemical_Process_Equipments (accessed on 22 September 2021).

52. Ziółkowski, P.; Kowalczyk, T.; Kornet, S.; Badur, J. On low-grade waste heat utilization from a supercritical steam power plant using an ORC-bottoming cycle coupled with two sources of heat. *Energy Convers. Manag.* **2017**, *146*, 158–173. [CrossRef]

53. Ziółkowski, P.; Kowalczyk, T.; Lemański, M.; Badur, J. On energy, exergy, and environmental aspects of a combined gas-steam cycle for heat and power generation undergoing a process of retrofitting by steam injection. *Energy Convers. Manag.* **2019**, *192*, 374–384. [CrossRef]

54. Werle, S.; Wilk, R.K. A review of methods for the thermal utilization of sewage sludge: The Polish perspective. *Renew. Energy* **2010**, *35*, 1914–1919. [CrossRef]

55. Schweitzer, D.; Gredinger, A.; Schmid, M.; Waizmann, G.; Beirow, M.; Spörl, R.; Scheffknecht, G. Steam gasification of wood pellets, sewage sludge and manure: Gasification performance and concentration of impurities. *Biomass Bioenergy* **2018**, *111*, 308–319. [CrossRef]

56. Akkache, S.; Hernández, A.-B.; Teixeira, G.; Gelix, F.; Roche, N.; Ferrasse, J.H. Co-gasification of wastewater sludge and different feedstock: Feasibility study. *Biomass Bioenergy* **2016**, *89*, 201–209. [CrossRef]

57. Bonalumi, D.; Valenti, G.; Lillia, S.; Fosbol, P.L.; Thomsen, K. A layout for the Carbon Capture with Aqueous Ammonia without Salt Precipitation. *Energy Procedia* **2016**, *86*, 134–143. [CrossRef]

58. Campanari, S.; Chiesa, P.; Manzolini, G. CO$_2$ capture from combined cycles integrated with Molten Carbonate Fuel Cells. *Int. J. Greenh. Gas Control* **2010**, *4*, 441–451. [CrossRef]

59. Gou, C.; Cai, R.; Hong, H. An Advanced Oxy-Fuel Power Cycle with High Efficiency. *Proc. Inst. Mech. Eng. Part A J. Power Energy* **2006**, *220*, 315–325. [CrossRef]

60. Liu, C.; Chen, G.; Sipöcz, N.; Assadi, M.; Bai, X. Characteristics of oxy-fuel combustion in gas turbines. *Appl. Energy* **2012**, *89*, 387–394. [CrossRef]

Article

Entrained Flow Plasma Gasification of Sewage Sludge–Proof-of-Concept and Fate of Inorganics

Vishwajeet [1], Halina Pawlak-Kruczek [1,*], Marcin Baranowski [1], Michał Czerep [1], Artur Chorążyczewski [2], Krystian Krochmalny [1], Michał Ostrycharczyk [1], Paweł Ziółkowski [3], Paweł Madejski [4], Tadeusz Mączka [5], Amit Arora [6], Tomasz Hardy [1], Lukasz Niedzwiecki [1], Janusz Badur [7] and Dariusz Mikielewicz [3]

[1] Department of Energy Conversion Engineering, Faculty of Mechanical and Power Engineering, Wroclaw University of Science and Technology, 50-370 Wrocław, Poland; vishwajeet.na@pwr.edu.pl (V.); marcin.baranowski@pwr.edu.pl (M.B.); michal.czerep@pwr.edu.pl (M.C.); krystian.krochmalny@pwr.edu.pl (K.K.); michal.ostrycharczyk@pwr.edu.pl (M.O.); tomasz.hardy@pwr.edu.pl (T.H.); lukasz.niedzwiecki@pwr.edu.pl (L.N.)
[2] Department of Automatic Control, Faculty of Electronics, Mechatronics and Control Systems, Wroclaw University of Science and Technology, 50-370 Wrocław, Poland; artur.chorazyczewski@pwr.edu.pl
[3] Faculty of Mechanical Engineering and Ship Technology, Institute of Energy, Gdańsk University of Technology, 80-233 Gdańsk, Poland; pawel.ziolkowski1@pg.edu.pl (P.Z.); dariusz.mikielewicz@pg.edu.pl (D.M.)
[4] Department of Power Systems and Environmental Protection Facilities, Faculty of Mechanical Engineering and Robotics, AGH University of Science and Technology, 30-059 Kraków, Poland; madejski@agh.edu.pl
[5] Institute of Power Systems Automation, 51-618 Wrocław, Poland; tadeusz.maczka@iase.wroc.pl
[6] Department of Chemical Engineering, Shaheed Bhagat Singh State University, Ferozepur 152004, Punjab, India; amitarora@sbsstc.ac.in or aroraamitlse@yahoo.com
[7] Energy Conversion Department, Institute of Fluid Flow Machinery, Polish Academy of Sciences, 80-231 Gdańsk, Poland; jb@imp.gda.pl
* Correspondence: halina.pawlak@pwr.edu.pl

Citation: Vishwajeet; Pawlak-Kruczek, H.; Baranowski, M.; Czerep, M.; Chorążyczewski, A.; Krochmalny, K.; Ostrycharczyk, M.; Ziółkowski, P.; Madejski, P.; Mączka, T.; et al. Entrained Flow Plasma Gasification of Sewage Sludge–Proof-of-Concept and Fate of Inorganics. *Energies* **2022**, *15*, 1948. https://doi.org/10.3390/en15051948

Academic Editor: Adam Smoliński

Received: 10 January 2022
Accepted: 3 March 2022
Published: 7 March 2022

Publisher's Note: MDPI stays neutral with regard to jurisdictional claims in published maps and institutional affiliations.

Abstract: Sewage sludge is a residue of wastewater processing that is biologically active and consists of water, organic matter, including dead and living pathogens, polycyclic aromatic hydrocarbons, and heavy metals, as well as organic and inorganic pollutants. Landfilling is on the decline, giving way to more environmentally friendly utilisation routes. This paper presents the results of a two-stage gasification–vitrification system, using a prototype-entrained flow plasma-assisted gasification reactor along with ex situ plasma vitrification. The results show that the use of plasma has a considerable influence on the quality of gas, with a higher heating value of dry gas exceeding 7.5 MJ/m_N^3, excluding nitrogen dilution. However, dilution from plasma gases becomes the main problem, giving a lower heating value of dry gas with the highest value being 5.36 MJ/m_N^3 when dilution by nitrogen from plasma torches is taken into account. An analysis of the residues showed a very low leaching inclination of ex-situ vitrified residues. This suggests that such a system could be used to avoid the problem of landfilling significant amounts of ash from sewage sludge incineration by turning inorganic residues into a by-product that has potential use as a construction aggregate.

Keywords: gasification; plasma; sewage sludge; inorganics; leaching

1. Introduction

Sewage sludge is a residue created during wastewater treatment and is becoming increasingly troublesome. Water, organic matter, as well as organic and inorganic contaminants, including polycyclic aromatic hydrocarbons (PAHs) and heavy metals, can be found in sewage sludge in various concentrations [1,2]. The possibility of biological activity in sewage sludge must be considered when it is used. As a result, a great deal of study has been done on sewage sludge deactivation and stabilisation [3–8] through many different thermal utilisation routes: thermal hydrolysis [9], hydrothermal carbonisation (HTC) [10–12], HTC integrated with anaerobic digestion [13,14], torrefaction [15,16], pelletising [17], pyrolysis and co-pyrolysis [18,19], gasification [20–23], and combustion [24].

Sewage sludge is regulated at both the European Union (EU) and national levels due to a variety of environmental, health, and safety issues. In the case of national rules, some are more stringent than the criteria laid down by EU legislation [25–27]. Figure 1, which is based on Eurostat's official statistics [28], shows the production of sewage sludge in all EU countries and the amounts currently applied in agriculture and incineration. There are approx. 60,000 wastewater treatment plants across Europe [29], and the location of plants might not always be logistically favourable when application in agriculture (land spreading) is considered, which is related to the high moisture content of the material. Agricultural application is also limited by the permissible limits on heavy metal content, i.e., 20 mg/kg$_{dry}$ of Cd, 1000 mg/kg$_{dry}$ of Cu, 16 mg/kg$_{dry}$ of Hg, 300 mg/kg$_{dry}$ of Ni, 750 mg/kg$_{dry}$ of Pb, and 2500 mg/kg$_{dry}$ of Zn, as specified in European Council Directive 86/278/EEC [30]. For agricultural use, the threat of microplastics should also be taken into serious consideration [31].

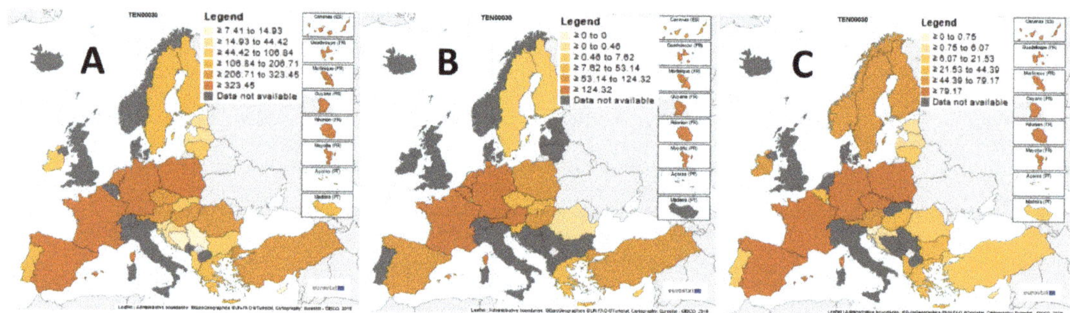

Figure 1. Sewage sludge production and disposal from urban wastewater (unit–thousand tonnes of dry mass) in 2016 based on the Eurostat Database [28]: (**A**) sewage sludge production (total); (**B**) sewage sludge utilisation by incineration; (**C**) sewage sludge disposal by agricultural use (data for later years not reported by many countries).

The following regulations are important on the EU level [26,27]:

- The Water Framework Directive (2000/60/EC)
- Directive 91/271/EEC (amended by Council Directive 2013/64/EU of 17 December 2013)
- Integrated pollution prevention and control directive (Directive 96/61/EC),
- Waste Landfilling Directive (99/31/EC)
- Sludge Use in Agriculture Directive (86/278/EEC)

Conversion of solid fuel to gas is the main point of gasification [32,33]. When air is used as a gasification agent, the main product called "producer gas" consists mainly of H_2, CO, CO_2, and N_2 [34–36]. Hydrocarbons are also present, among which methane is the most significant of all non-condensable gases [37,38]. Phenols, toluene, naphthalene, benzene, and other aromatic chemicals, as well as more complex condensable compounds, are also present [39–42]. Compounds with an atomic mass higher than benzene are often referred to as tars [43–46]. Gasification has been investigated extensively for many different materials [47–54].

Intensive investigation into the gasification of sewage sludge and the subsequent use of producer gas has been performed for many years. Werle discovered that the laminar flame speed rose along with the hydrogen level in the producer gas from sewage sludge [55]. Such gas could be usedin spark-ignition engines [56]. Nonetheless, to get adequate performance out of a spark-ignition engine, producer gas from sewage sludge requires a 40% addition of methane, according to Szwaja et al. [57].

According to Werle and Dudziak, phenols and their derivatives make up the great bulk of tars produced by gasification of sewage sludge [58]. Pawlak-Kruczek et al. [59] recommended using a tar-deposition diagram to predict the potential severity of tar deposition problems in gas coolers. Using this tool, the study demonstrated that significant torrefaction of sewage sludge prior to steam gasification reduced the content of tars with melting points above 40 °C [59]. In terms of sewage sludge thermal utilisation, gasification has been considered as an interesting alternative to incineration, with some works even proposing such a thermal route, leading to gas-powered plants with negative CO_2 emissions [60].

The amount of information available in the literature on commercial sewage sludge gasification plants is quite minimal. A case study on the commercial-scale gasification of sewage sludge for the Greek island of Psittaleia was conducted using the Gasif Eq equilibrium model, which was performed by Montouris et al. [61]. It was discovered that plasma gasification of sewage sludge could result in net power production [61]. The calculation, which was carried out for a hypothetical plant with a processing rate of 250 tonnes per day (moisture content 68%), revealed the possibility of supplying 2.85 MW of electricity [61]. Experiments using two-step plasma processing units were effective for several research groups [62,63], proving the concept's general practicality in a lab setting and the prospect of lowering the tar level to 90 mg/m$_N^3$ [63]. Brachi et al. [64] determined that gasification of sewage sludge, with the combustion of the gas in a CHP unit, could be economically feasible for a real wastewater treatment plant that serves a 1.2 million population equivalent in Southern Italy.

2. Aim of the Study and Justification

State-of-the-art incineration of sewage sludge is capable of significantly reducing its mass. However, the content of incombustible inorganics (ash) in sewage sludge may reach a value as high as a quarter or even a third of its dry mass (e.g., see Table 1). Moreover, combustion is never complete, which adds to the total mass of the waste still left after combustion. Therefore, it is plausible to state that incineration is only a partial solution to the sewage sludge problem because a significant part of the mass of the original waste stream still needs to be landfilled.

Table 1. Range of values for proximate and ultimate analysis of sewage sludge, based on our own analyses using samples from wastewater treatment plants in Wrocław (Janówek) and Brzeg (all data given on dry basis).

Proximate Analysis	
Volatile Matter content	56.0–58.1%$_{dry}$
Fixed Carbon	9.4–17.8%$_{dry}$
Ash content	26.2–32.5%$_{dry}$
Higher Heating Value	13.66–15.70 MJ/kg
Ultimate Analysis	
C content	27.89–32.16%$_{dry}$
H content	2.86–6.67%$_{dry}$
N content	4.36–4.83%$_{dry}$
S content	0.29–0.81%$_{dry}$
O content	28.80–33.14%$_{dry}$

After incineration, both bottom and fly ash have their own respective waste codes, according to the European Parliament and Council Regulation (EC) No 2150/2002 on waste statistics, enacted on 25 November 2002. Hazardous and non-hazardous bottom ash and slag receive the waste codes 19 01 11 and 19 01 12, respectively. Similarly, fly ash receives the code 19 01 13 or 19 01 14, depending on the hazard involved. Such waste is then deemed ready to be directed to the appropriate landfill which is selected based on Council Decision 2003/33/EC of 19 December 2002, which establishing procedures and criteria for accepting waste at landfills. The document states specific limits regarding leaching limits for different types of landfills. This implies a certain cost, i.e., gate fees associated with landfilling along with the transportation cost of the waste to an appropriate landfill site.

Such cost can be avoided if the waste is turned into a product, which is possible based on Directive 2008/98/EC of the European Parliament and of the Council of 19 November 2008. In Article 6, the directive outlines the requirements for obtaining end-of-waste status: waste which has undergone a recycling or other recovery operation is considered to have ceased to be waste if it complies with the following conditions:

- the substance or object is to be used for specific purposes,
- a market or demand exists for such a substance or object,
- the substance or object fulfils the technical requirements for the specific purposes and meets the existing legislation and standards applicable to products, and
- the use of the substance or object will not lead to overall adverse environmental or human health impacts.

Moreover, promoting such an approach is beneficial for member states because turning sewage sludge into fuel is not counted towards the attainment of the recycling targets, as stated by the directive: (Article 11a–Rules on the calculation of the attainment of the targets).

The outcome would not be much different for technologies competitive with incineration, such as gasification, in which unconverted carbon would also significantly contribute to the amount of waste left after the process. Therefore, in the case of gasification, the possibility to turn post-process solid waste into a product should not be overlooked. However, such a product would still need to comply with the requirements regarding avoiding adverse environmental or human health impacts and the existing market for such product. Environmental and human health impacts could be effectively minimised if the waste is inert, which could be determined based on requirements set by the Council Decision 2003/33/EC of 19 December 2002 (see Table 2).

Table 2. Leaching limit values for inert waste (for different allowed test procedures) set by the Council Decision 2003/33/EC of 19 December 2002 [65] (dry substance—data given on dry basis).

Component	Test		
	L/S = 2 dm^3/kg	L/S = 10 dm^3/kg	C0 (Percolation Test)
	mg/kg dry substance	mg/kg dry substance	mg/dm^3
As	0.1	0.5	0.06
Ba	7	20	4
Cd	0.03	0.04	0.02
Cr total	0.2	0.5	0.1
Cu	0.9	2	0.6
Hg	0.003	0.01	0.002
Mo	0.3	0.5	0.2
Ni	0.2	0.4	0.12
Pb	0.2	0.5	0.15
Sb	0.02	0.06	0.1
Se	0.06	0.1	0.04
Zn	2	4	1.2
Chloride	550	800	460
Fluoride	4	10	2.5
Sulphate	560	1 000	1 500
Phenol index	0.5	1	0.3

One of the promising ways to obtain such values is by vitrification, which changes the structure of waste, such as ash, in a way that makes its structure resemble amorphous crystals, such as glass. Such a structure could effectively immobilise hazardous waste components, such as heavy metals, and significantly decrease the rate of leaching, thus enabling the possibility of such waste to be considered inert.

The aim of this study is to provide a proof-of-concept for the two-step plasma gasification of sewage sludge with vitrification of inorganic residues, thereby allowing the residue to be turned into a valuable product that complies with all norms regarding its influence on the environment.

3. Materials and Methods

The gasification of sewage sludge was performed using a bespoke rig as shown in Figure 2. The rig was equipped with a plasma torch, using N_2 to generate the plasma. The reactor wall was built of stainless steel with a ceramic refractory. The temperature at the edge of the refractory was measured using a K-type thermocouple inserted into the top revisory hole. Gasification in the reactor was performed in an entrained flow, with residence time on the order of magnitude of 1 s. Pre-dried sewage sludge was brought from the Municipal Wastewater Treatment plant in Janówek (near Wrocław, Poland). The plant in Janówek is equipped with anaerobic digestion reactors. After anaerobic digestion, the sludge is dried in a rotary drum dryer. Samples of pre-dried sewage sludge were taken at the outlet of the drier.

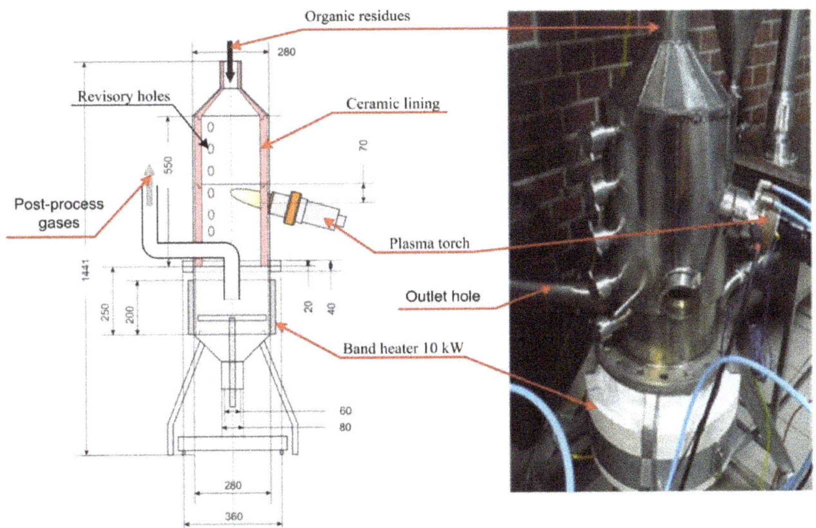

Figure 2. Experimental rig for plasma gasification of sewage sludge.

An experimental matrix for performed entrained flow plasma-assisted gasification experiments is provided in Table 3. Gasification was performed with different plasma-gas-to-fuel ratios (PGFR) as well as different air-to-fuel ratios (AFR). Mass flow rates of plasma gas were calculated using flow rates measured with rotameters and density of nitrogen in normal conditions. Additionally, during gasification, air was supplied to the reactor along with fine particles of sewage sludge (d < 1 mm), entrained from the auger located on the top of the gasifier. The two-phase mixture was fed from the top of the reactor into the freeboard, located over the streams of hot plasma, generated by plasma torches. The producer gas was removed from the bottom of the reactor (see Figure 2). Tangential placements of plasma torches enforced cyclonic flow inside of the reactor.

Table 3. Experimental matrix for plasma-assisted gasification of pre-dried sewage sludge (R-wall indicates refractory wall).

Exp. ID	AFR	PGFR	I	$T_{R\text{-wall}}$
	–	–	A	°C
I	0.43	6.73	100	650
II	0.33	7.36	80	620
III	0.29	6.73	100	680
IV	0.33	7.85	100	680
V	0.29	6.31	80	620

Ex situ vitrification experiments were performed using a plasma torch located over a pile of post-gasification residues in a simple rectangular reactor, built using heat resistant bricks. Post-gasification residues, gathered during gasification experiments, sintered and melted, whereas any additional gases created during the process were directed into the fume hood and subsequently to the ventilation system. The vitrified residue was subsequently removed from the reactor using a chisel. Furthermore, a chisel was used to chip off any small fragments of brick lining from the vitrified sample. Subsequently, leaching tests were performed in an external laboratory, in compliance with methods outlined by the EU in the Council Decision 2003/33/EC of 19 December 2002.

Proximate analysis was performed using a Perkin–Elmer Diamond TGA (thermogravimetric analyser). The following program was applied during tests:

➤ Heat to 105 °C; ramp at 10 °C/min + hold for 10 min
➤ (2 a) Air was used to determine ash content: Heat to 815 °C; ramp at 50 °C/min + hold for 15 min
➤ (2 b) N_2 was used to determine the volatile matter content: Heat to 850 °C; ramp at 50 °C/min + hold for 15 min

The IKA C2000 basic bomb calorimeter was used to calculate the higher heating value, in compliance with ISO 1928. The isoperibolic method was used. Ultimate analysis was performed using Perkin–Elmer 2400 analyser, according to polish standard PKN-ISO/TS 12902:2007.

Oxide analysis was performed using the Atomic Absorption Spectrometry method, with AAnalyst 400 analyser. Samples were burned in the oven under the ashing temperature equal to 815 °C (residence time–3 h). Afterwards, between 100 and 150 mg of ash were diluted in 5 mL of HNO_3 and 3 mL of HF and subsequently mineralized in a Multiwave 3000 microwave oven, under 250 °C and 60 bar (pressure ramp 0.5 bar/s) for 80 min. Mineralized samples were diluted in 18 mL of saturated boric acid to bind free fluorides. Then the solution was diluted using distilled water (18.2 MΩ·cm) to obtain the final sample volume of 100 mL.

4. Results and Discussion

4.1. Two-Stage Sewage Sludge Utilisation Process–Stage I: Plasma-Assisted Gasification

The pre-dried sewage sludge from the wastewater treatment plant in Janówek, near Wrocław, was characterised, and the results of the proximate and ultimate analysis of the gasification feedstock are reported in Table 4.

The results for each of the plasma-assisted gasification experiments are shown in Table 4 and Figures 3–7. It can be clearly seen that the dilution strongly influenced the lower heating values of the producer gas (Table 5). However, results of calculations performed for the dry producer gas, without taking inert nitrogen into account, showed that the use of plasma positively influenced the heating value of the producer gas since values of HHV, close to 8 MJ/m_N^3 could be achieved. The real LHV values (Table 5) are similar to those obtained by Striūgas et al. [63], who performed plasma-assisted gasification of sewage sludge and achieved an LHV of 4.82 MJ/m_N^3.

Table 4. Proximate and ultimate analysis of the pre-dried sewage sludge from Municipal Wastewater Treatment plant in Janówek (near Wrocław, Poland): HHV–Higher Heating Value; C–carbon content; H–hydrogen content; N–nitrogen content; S–sulphur content; O–oxygen content] (dry—data given on dry basis; as received—data given on as received basis).

	Value	Unit
Volatile matter	56.0	$\%_{dry}$
Fixed carbon	17.8	$\%_{dry}$
Ash	26.2	$\%_{dry}$
Moisture	7.5	$\%_{as\ received}$
HHV	13.658	MJ/kg
C	32.16	$\%_{dry}$
H	2.86	$\%_{dry}$
N	4.83	$\%_{dry}$
S	0.81	$\%_{dry}$
O	33.14	$\%_{dry}$

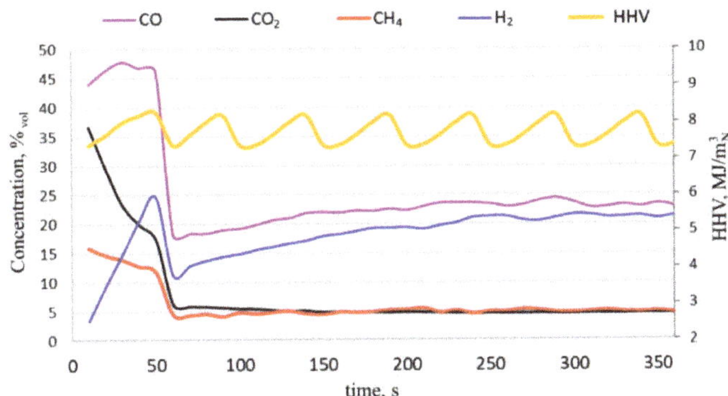

Figure 3. Plasma assisted gasification of sewage sludge–ID I (concentrations after excluding nitrogen supplied by plasma torch).

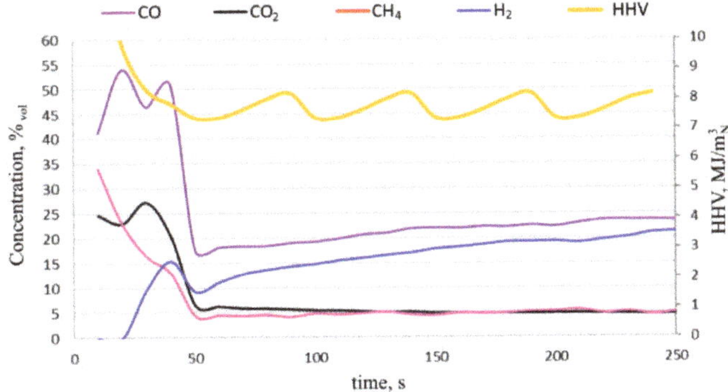

Figure 4. Plasma assisted gasification of sewage sludge–ID II (concentrations after excluding nitrogen supplied by plasma torch).

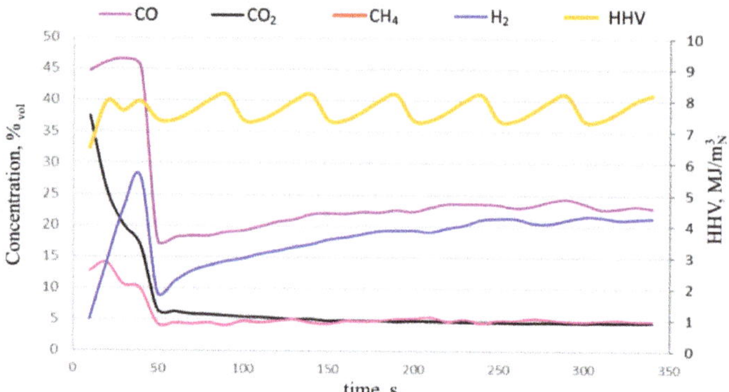

Figure 5. Plasma assisted gasification of sewage sludge–ID III (concentrations after excluding nitrogen supplied by plasma torch).

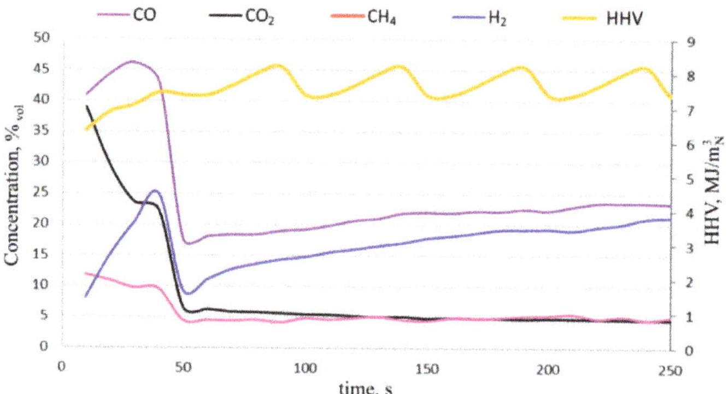

Figure 6. Plasma assisted gasification of sewage sludge–ID IV (concentrations after excluding nitrogen supplied by plasma torch).

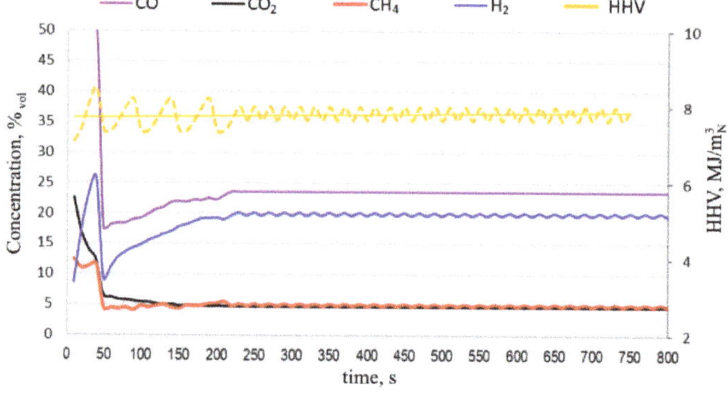

Figure 7. Plasma assisted gasification of sewage sludge–ID V (concentrations after excluding nitrogen supplied by plasma torch).

Table 5. Lower heating value (LHV) of dry producer gas for each test.

Experiment ID	Average LHV of the Gas, MJ/m_N^3
I	4.80
II	4.00
III	4.70
IV	3.96
V	5.36

Temperature is important in the gasification process, which is also the case for sewage sludge gasification [66]. Werle noticed a decrease in temperature and an increase in the concentration of combustible components in the producer gas with an increased oxygen content of the sludge [67]. Moreover, increased air temperature at the entrance of a fixed bed gasifier, according to observation, enhanced the production of combustible chemicals during the gasification of sewage sludge [68]. However, the temperature of the walls of the reactor was between 620 and 680 °C, which suggested that the average temperature in the reactor was much smaller than for Striūgas et al. [63], where the temperature at the exit was 1100 °C [63]. The hydrogen content in the gas was higher in comparison to the work of Striūgas et al. [63], reaching $20\%_{vol}$ compared to approx. $14.5\%_{vol}$ [63]. This was much less than in the case of fluidised bed gasification and fixed bed gasification, where observed concentrations of hydrogen were higher than 40% [69] and 30% [70], respectively. This suggested that the plasma treatment slightly decreased hydrogen content in the treated producer gas. This was in qualitative agreement with the results of Wnukowski et al. [71], who reported treatment of a model (artificially prepared) producer gas with microwave plasma and observed a slight decrease in H_2 content [71]. In the work of Striūgas et al. [63], plasma was applied to gas from a downdraft gasifier in a separate reactor. In the work of Chun and Song, microwave-induced fixed gasification of sewage sludge was performed, also giving an H_2 content close to $20\%_{vol}$ [72]. This suggested that the presence of gasified sewage sludge particles in the plasma also has some influence on the composition of the gas and on reaction pathways. This could suggest some autocatalytic effect and requires further investigation.

In each of the cases, the startup took only about 50 s. After the startup, an initial stabilisation period was observed when the gasifier was not fully in equilibrium and conditions were gradually changing, with the heating value of the gas changing in a cycle. Such cycles can be explained by changes in C_xH_y in the composition of the gas, caused by pyrolysis of relatively heavy sewage sludge that was falling at the bottom of the reactor. After devolatilisation, the particles of char from sewage sludge subsequently became lighter and were entrained. Such a stabilisation period took an additional 200 s, depending on the temperature (Figures 3–7). It was followed by a steady-state period, which was distinguished by the decreased amplitude of cycling the HHV of syngas. The longest test was performed for 13 min (Figure 7), so it could reasonably be expected that long-term stable operation of the unit is possible, provided the variability in the fuel quality is not high and the wear on the plasma torches is not significant.

4.2. Two-Stage Sewage Sludge Utilisation Process—Stage II: Plasma Vitrification

In general, vitrification aimed at obtaining an aggregate that could be used as a construction material can be performed in two different ways: in situ and ex situ. Ex situ vitrification can be performed for the residue after gasification, as well as for the ash after incineration. Ex situ vitrification has some disadvantages, as unburned carbon can still be converted during this process by oxidation, which could result in gaseous by-products, which would contribute to emissions of the installation. Conversion could be avoided with the introduction of an inert atmosphere for vitrification. However, in such a case, a high cost would be involved with the production or purchase of nitrogen. Furthermore, this would mean an additional unit operation in the process chain. On the other hand, in situ vitrification can be applied for gasification inside the reactor if temperatures are sufficiently high. This is possible if sufficiently high amounts of heat are provided to the gasification reactor, which could be achieved using plasma torches. Benefits include high carbon conversion and stabilisation of the process, which offers better controllability, making it less sensitive to poor-quality fuels. Such an effect could also be achieved by using oxygen or air pre-heated to high temperatures as a gasifying agent.

The results (Table 6) showed that vitrified residues could be considered inert material. Only the content of phenols exceeded the limits. This was probably a result of soot depositions created from residual carbon, but overall, this can be amended by improvements in the offtake of gases from vitrification.

Table 6. Leaching for vitrified residues after plasma gasification of sewage sludge from the wastewater treatment plant in Wrocław–Janówek (leaching performed with L/S = 10 L/kg; dry substance—data given on dry basis).

Component	Obtained Value	Limit for Inert Materials
	mg/kg dry substance	mg/kg dry substance
As	<0.10	0.5
Ba	0.09 ± 0.03	20
Cd	<0.01	0.04
Cr total	<0.01	0.5
Cu	0.050 ± 0.016	2
Hg	<0.0001	0.01
Mo	<0.20	0.5
Ni	<0.01	0.4
Pb	<0.01	0.5
Sb	<0.20	0.06
Se	<0.20	0.1
Zn	<0.01	4
Chloride	<10.0	800
Fluoride	<1.00	10
Sulphate	394 ± 118	1 000
Phenol index	54 ± 12	1

It could be noticed that the composition of ash in post-gasification residues did not deviate significantly with respect to raw sewage sludge (see Table 7). The temperature of the reactor did not seem to have a significant influence on the ash composition as only two tests deviated significantly from the rest, namely ID II and ID IV. The temperature of the reactor during ID II was relatively low, looking at the temperature of the edge of the wall (see Table 4), whereas the temperature during ID IV was among the highest. During ID II, the concentration of Al_2O_3 was slightly higher, whereas for ID IV the concentration of SiO_2 was slightly higher compared to other tests. Moreover, for ID IV, the concentration of CaO was slightly lower than in other tests, as well as for the raw sewage sludge sample. The inhomogeneous character of sewage sludge seems to be the most plausible explanation for such behaviour.

Table 7. Ash composition for residues from different gasification trials.

Sample	K_2O	Na_2O	CaO	Mn_3O_4	Fe_2O_3	MgO	Al_2O_3	TiO_2	SiO_2
ID I	1.71	0.20	26.7	0.19	16.1	3.43	7.84	1.30	42.57
ID II	1.88	0.85	24.6	0.19	15.7	2.70	12.00	1.32	40.72
ID III	1.74	0.76	25.3	0.18	15.7	2.77	6.80	1.27	45.43
ID IV	1.57	0.93	19.0	0.17	14.5	3.16	7.97	1.10	51.55
ID V	1.54	0.67	25.5	0.19	15.3	2.71	7.86	1.28	44.98
Raw sewage sludge	1.63	0.44	28.2	0.18	16.4	3.46	6.99	1.22	41.45

5. Conclusions

Overall, the results of the leaching experiments using ex situ vitrified post-gasification residues, showed that the two-stage plasma gasification system can be efficiently used to make the inorganic part of sewage sludge inert. This showed that such a 2-stage system could be used for the thermal treatment of sewage sludge, allowing its use for energy purposes, with a possibility of the inorganic part reaching end-of-waste status, thus becoming a marketable product. Thus, the work could be considered a proof-of-concept. Nonetheless, the results of phenol leaching indicated that the off-gases from the second stage of the process (vitrification) could be problematic. Therefore, it seems plausible to suspect that a single-stage system with in situ vitrification could bring some additional benefits. Alternatively, a combination of non-plasma gasification and plasma vitrification would consume less electricity and still offer the benefits of fulfilling an end-of-waste protocol for inorganic residues and turning them into a marketable product. Plasma gasification makes sense only when the plasma-generating agent is not inert, which has a positive influence on the heating value of the gas without diluting it. Furthermore, additional valorisation, e.g., hydrothermal carbonization, and its influence on plasma gasification is a promising area for future studies. Moreover, additional work is recommended to prove the feasibility of the concept from an economic standpoint.

Author Contributions: Conceptualisation: H.P.-K., T.M.; methodology: H.P.-K., M.B., A.C., T.M.; validation: H.P.-K., T.M., P.M., P.Z.; formal analysis: H.P.-K., M.B., A.C., P.Z.; investigation: M.B., M.C., A.C., K.K.; resources: D.M., H.P.-K., J.B., P.Z.; data curation: V., H.P.-K., M.B., M.O.; writing—original draft preparation: V., L.N.; writing—review and editing: V., L.N., A.A., T.H., D.M.; visualisation: T.M., P.M., P.Z., M.O.; supervision: D.M., H.P.-K., A.A., T.H.; project administration: D.M., H.P.-K., J.B., P.Z.; funding acquisition: P.Z., D.M., H.P.-K., J.B. All authors have read and agreed to the published version of the manuscript.

Funding: The research leading to these results has received funding from the Norway Grants 2014–2021 via the National Centre for Research and Development. The article has been prepared within the frame of the project: "Negative CO2 emission gas power plant"-NOR/POLNORCCS/NEGATIVE-CO2-PP/0009/2019-00, which is co-financed by program "Applied research" under the Norwegian Financial Mechanisms 2014–2021 POLNOR 2019–Development of CO_2 capture solutions integrated in power and industry processes.

Institutional Review Board Statement: Not applicable.

Informed Consent Statement: Not applicable.

Data Availability Statement: All the data is in the manuscript.

Acknowledgments: Authors would like to thank MPWiK Wrocław for supplying dried sewage sludge samples for this research.

Conflicts of Interest: The authors declare no conflict of interest. The funders had no role in the design of the study; in the collection, analyses, or interpretation of data; in the writing of the manuscript, or in the decision to publish the results.

References

1. Kacprzak, M.; Neczaj, E.; Fijałkowski, K.; Grobelak, A.; Grosser, A.; Worwag, M.; Rorat, A.; Brattebo, H.; Almås, Å.; Singh, B.R. Sewage sludge disposal strategies for sustainable development. *Environ. Res.* **2017**, *156*, 39–46. [CrossRef]
2. Werle, S. Sewage Sludge-To-Energy Management in Eastern Europe: A Polish Perspective. *Ecol. Chem. Eng. S* **2015**, *22*, 459–469. [CrossRef]
3. Trinh, T.T.; Werle, S.; Tran, K.-Q.; Magdziarz, A.; Sobek, S.; Pogrzeba, M. Energy crops for sustainable phytoremediation—Thermal decomposition kinetics. *Energy Proc.* **2019**, *158*, 873–878. [CrossRef]
4. Zubrowska-Sudol, M.; Walczak, J. Enhancing combined biological nitrogen and phosphorus removal from wastewater by applying mechanically disintegrated excess sludge. *Water Res.* **2015**, *76*, 10–18. [CrossRef]
5. Szatyłowicz, E.; Walczak, J.; Żubrowska-Sudoł, M.; Garlicka, A. Deactivation of the Activated Sludge As a Result of Mechanical Disintegration. *Inżynieria Ekol.* **2017**, *18*, 114–121. [CrossRef]
6. Zubrowska-Sudol, M.; Walczak, J. Effects of mechanical disintegration of activated sludge on the activity of nitrifying and denitrifying bacteria and phosphorus accumulating organisms. *Water Res.* **2014**, *61*, 200–209. [CrossRef]

7.   Żubrowska-Sudoł, M.; Podedworna, J.; Bisak, A.; Sytek-Szmeichel, K.; Krawczyk, P.; Garlicka, A. Intensification of anaerobic digestion efficiency with use of mechanical excess sludge disintegration in the context of increased energy production in wastewater treatment plants. *E3S Web Conf.* **2017**, *22*, 00208. [CrossRef]

8.   Szatyłowicz, E.; Garlicka, A.; Żubrowska-Sudoł, M. The Effectiveness of the Organic Compounds Released Due To the Hydrodynamic Disintegration of Sewage Sludge. *Inżynieria Ekol.* **2017**, *18*, 47–55. [CrossRef]

9.   Mukawa, J.; Pająk, T.; Rzepecki, T. Sewage sludge treatment in the process of thermal hydrolysis and digestion on the Tarnow sewage treatment plant example. *Przem. Chem.* **2018**, *1*, 92–94. [CrossRef]

10.   Czerwińska, K.; Śliz, M.; Wilk, M. Hydrothermal carbonization process: Fundamentals, main parameter characteristics and possible applications including an effective method of SARS-CoV-2 mitigation in sewage sludge. A review. *Renew. Sustain. Energy Rev.* **2022**, *154*, 111873. [CrossRef]

11.   Aragón-Briceño, C.I.; Grasham, O.; Ross, A.B.; Dupont, V.; Camargo-Valero, M.A. Hydrothermal carbonization of sewage digestate at wastewater treatment works: Influence of solid loading on characteristics of hydrochar, process water and plant energetics. *Renew. Energy* **2020**, *157*, 959–973. [CrossRef]

12.   Wilk, M.; Magdziarz, A.; Jayaraman, K.; Szymańska-Chargot, M.; Gökalp, I. Hydrothermal carbonization characteristics of sewage sludge and lignocellulosic biomass. A comparative study. *Biomass Bioenergy* **2019**, *120*, 166–175. [CrossRef]

13.   Aragón-Briceño, C.; Ross, A.B.B.; Camargo-Valero, M.A.A. Evaluation and comparison of product yields and bio-methane potential in sewage digestate following hydrothermal treatment. *Appl. Energy* **2017**, *208*, 1357–1369. [CrossRef]

14.   Aragón-Briceño, C.I.; Ross, A.B.; Camargo-Valero, M.A. Mass and energy integration study of hydrothermal carbonization with anaerobic digestion of sewage sludge. *Renew. Energy* **2021**, *167*, 473–483. [CrossRef]

15.   Pawlak-Kruczek, H.; Krochmalny, K.K.; Wnukowski, M.; Niedzwiecki, L. Slow pyrolysis of the sewage sludge with additives: Calcium oxide and lignite. *J. Energy Resour. Technol.* **2018**, *140*, 062206. [CrossRef]

16.   Pawlak-Kruczek, H.; Wnukowski, M.; Krochmalny, K.; Kowal, M.; Baranowski, M.; Zgóra, J.; Czerep, M.; Ostrycharczyk, M.; Niedzwiecki, L. The Staged Thermal Conversion of Sewage Sludge in the Presence of Oxygen. *J. Energy Resour. Technol.* **2019**, *141*, 070701. [CrossRef]

17.   Nagy, D.; Balogh, P.; Gabnai, Z.; Popp, J.; Oláh, J.; Bai, A. Economic Analysis of Pellet Production in Co-Digestion Biogas Plants. *Energies* **2018**, *11*, 1135. [CrossRef]

18.   Wang, S.; Persson, H.; Yang, W.; Jönsson, P.G. Pyrolysis study of hydrothermal carbonization-treated digested sewage sludge using a Py-GC/MS and a bench-scale pyrolyzer. *Fuel* **2020**, *262*, 116335. [CrossRef]

19.   Wang, S.; Mandfloen, P.; Jönsson, P.; Yang, W. Synergistic effects in the copyrolysis of municipal sewage sludge digestate and salix: Reaction mechanism, product characterization and char stability. *Appl. Energy* **2021**, *289*, 116687. [CrossRef]

20.   Dudziak, M.; Werle, S.; Grübel, K. Evaluation of Contamination of Dried Sewage Sludge and Solid By-Products of Dried Sewage Sludge Gasification By Infrared Spectroscopy Method. *Inżynieria Ekol.* **2016**, *50*, 195–200. [CrossRef]

21.   Werle, S. Gasification of a Dried Sewage Sludge in a Laboratory Scale Fixed Bed Reactor. *Energy Proc.* **2015**, *66*, 253–256. [CrossRef]

22.   Werle, S.; Dudziak, M. Pollution of liquid waste-products from sewage sludge gasification. *Proc. ECOpole* **2015**, *9*, 15–17. [CrossRef]

23.   Werle, S. Gasification of a Dried Sewage Sludge in a Laboratory Scale Fixed Bed Reactor. *Energies* **2015**, *8*, 8562–8572. [CrossRef]

24.   Pająk, T. Thermal treatment as sustainable sewage sludge management. *Environ. Prot. Eng.* **2013**, *39*, 41–53. [CrossRef]

25.   Andersen, A. *Disposal and Recycling Routes for Sewage Sludge Part 2: Regulatory Report*; Office for Official Publications of the European Communities: Luxembourg, 2001.

26.   Inglezakis, V.J.; Zorpas, A.A.; Karagiannidis, A.; Samaras, P.; Voukkali, I.; Sklari, S. European Union Legislation on Sewage Sludge Management. *Fresenius Environ. Bull.* **2014**, *23*, 635–639.

27.   Inglezakis, V.; Zorpas, A.A.; Karagiannidis, A.; Samaras, P.; Voukkali, I. European Union legislation on sewage sludge management. In Proceedings of the 3rd International CEMEPE & SECOTOX Conference, Skiathos, Greece, 19–24 June 2011; pp. 475–480.

28.   Eurostat. Available online: https://ec.europa.eu/eurostat/databrowser/view/ten00030/default/map?lang=en (accessed on 26 February 2022).

29.   Bai, A.; Gabnai, Z. Opportunities of Circular Economy in a Complex System of Woody Biomass and Municipal Sewage Plants. In *Forest Biomass—From Trees to Energy*; IntechOpen: London, UK, 2021.

30.   Khakbaz, A.; De Nobili, M.; Mainardis, M.; Contin, M.; Aneggi, E.; Mattiussi, M.; Cabras, I.; Busut, M.; Goi, D. Monitoring of heavy metals, eox and las in sewage sludge for agricultural use: A case study. *Detritus* **2020**, *12*, 160–168. [CrossRef]

31.   Milojevic, N.; Cydzik-Kwiatkowska, A. Agricultural Use of Sewage Sludge as a Threat of Microplastic (MP) Spread in the Environment and the Role of Governance. *Energies* **2021**, *14*, 6293. [CrossRef]

32.   Mularski, J.; Pawlak-Kruczek, H.; Modlinski, N. A review of recent studies of the CFD modelling of coal gasification in entrained flow gasifiers, covering devolatilization, gas-phase reactions, surface reactions, models and kinetics. *Fuel* **2020**, *271*, 117620. [CrossRef]

33.   Wang, Z.; Li, J.; Burra, K.G.; Liu, X.; Li, X.; Zhang, M.; Lei, T.; Gupta, A.K. Synergetic Effect on CO_2-Assisted Co-Gasification of Biomass and Plastics. *J. Energy Resour. Technol.* **2021**, *143*, 031901. [CrossRef]

34.   Čespiva, J.; Skřínský, J.; Vereš, J.; Borovec, K.; Wnukowski, M. Solid-recovered fuel to liquid conversion using fixed bed gasification technology and a fischer–tropsch synthesis unit—Case study. *Int. J. Energy Prod. Manag.* **2020**, *5*, 212–222. [CrossRef]

- Oxy-combustion carbon capture occurs after the combustion process in an oxygen atmosphere by separating CO2 generated during the oxy-combustion process, e.g., using an oxygen gas turbine. Oxygen atmosphere can be obtained by removing nitrogen from the air before the combustion process.

A diagram explaining the methods and techniques for CO_2 capture is shown in Figure 1.

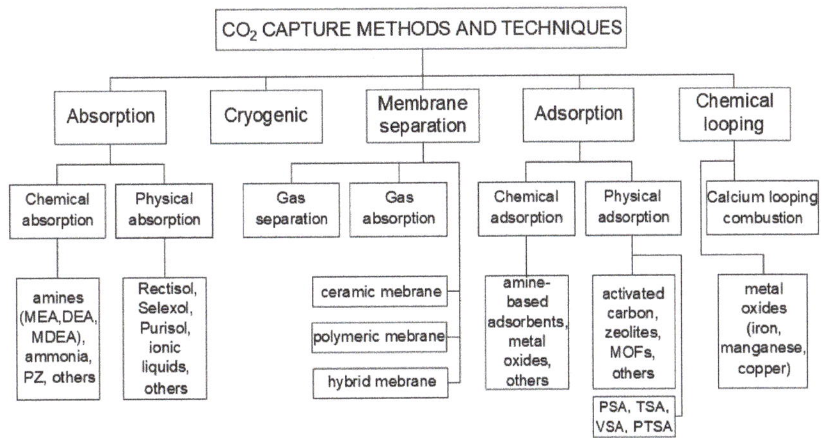

Figure 1. Methods and techniques of CO_2 capture.

In pre-combustion carbon capture, fuel is oxidized using a gasification process, which produces syngas with a composition of hydrogen (H_2) and carbon monoxide (CO). The produced CO is converted into CO_2, which is captured before combustion. Shijaz et al. showed a comparison between power generation using coal gasification without carbon capture, pre-combustion carbon capture, or chemical looping combat (CLC). The results showed that the overall efficiency of a power plant with pre-combustion carbon capture and CLC is reduced compared to a plant without CO_2 capture. CO_2 captured from fuel reduces the fuel volume sent to the turbine, reducing power generation. However, due to environmental concerns, including a CO_2 capture unit is unavoidable [13]. Mukherjee et al. compared each type of carbon capture method using an IGCC power plant without CO_2 capture. CLC was combined with a coal-fired IGCC, and analyses were performed based on electrical efficiency and carbon capture efficiency. The CLC and oxyfuel combustion methods showed a value of around a 100% carbon capture rate compared to pre-combustion CO_2 capture, which achieved 94.8% CO_2 capture. Another method of coal direct chemical looping combustion, where coal is fed directly to a boiler without gasification, increased the electrical efficiency and achieved 100% CO_2 capture. The results from the comparison of IGCC-CLC, pre-combustion, and oxyfuel combustion showed that the methods' energy penalties were 4.5%, 7.1%, and 9.1%, respectively [14]. This article describes the various carbon capture and storage methods and technology used in large-scale units. The CO_2 emission from a coal-fired ultra-supercritical power plant is calculated.

The SO_2 and NO_x content in the flue gas has a higher chance of affecting the purity of CO_2 during CO_2 capture. The high purity of the CO_2 stream is very important for recycling methods. A pilot, dual-reflux VPSA unit, installed in the Łagisza Power plant in Poland, was installed for post-combustion CO_2 capture. Before the flue gas from the boiler is sent to the DR-VPSA unit, the flue gas is passed through an absorber, an adsorber, and a glycolic gas dehydration system to remove SO_2 and NO_x. Activated carbon works effectively for the removal of SO_2 compared to the removal of NO_x, which leads to a high-purity CO_2 [15].

CO_2 emissions from the power sector are mainly caused by modern technologies, such as coal-fired, gas-fired, oil-fired, and combined cycle gas turbine (CCGT) power plants. Table 1 presents the CO_2 emissions and lower heating value depending on fossil fuel. From Table 1, it can be seen that CO_2 emissions depend on the content of the fuel. The higher the fuel content, the higher the CO_2 emissions will be. The increasing share of H_2 gives better properties to fossil fuels, taking into account the LHV and CO_2 emission levels. Natural gas consists mainly of CH_4 and is characterized by almost two times lower emissions than hard and lignite coal, and an almost two times larger LHV. This fact comes from gas fuel composition, where four atoms of hydrogen are inside the methane molecule of every carbon atom. Other modern power generating technologies, such as nuclear, renewable energy sources (RES), and hydrogen-based technologies, are less likely to produce emissions. A newly built CCGT power plant has CO_2 emissions of 350 kg$_{CO2}$/MWh without carbon capture, indicating the same emissions as a gas-fired power plant [16]. In the case of an ultra-supercritical power plant fired with coal, CO_2 emissions up to 700 kgCO$_2$/MWh can be produced. According to the type of coal-fired critical power plant used, CO_2 emissions range from 690 to 830 kgCO$_2$/MWh [17]. The use of fossil fuel in modern energy technology will continue until it is replaced with alternative technologies, such as renewable energies and power production without emission. Until these technologies are replaced with those without CO_2 emissions, CO_2 capture is unavoidable to reduce greenhouse gases and protect the environment.

Table 1. Emission of carbon dioxide (CO_2) during combustion of different hydrocarbon fuels [1–3,8].

Fuel	LHV	Emission
	MJ/kg	kgCO$_2$/GJ
Hard coal	>23.9	94.60
Lignite	<17.4	101.20
Crude oil	43.0	74.07
Petrol	43.4	66.00
Paraffin oil	41.5	71.50
Heating oil	42.8	77.37
Diesel	42.6	74.07
Natural gas	47.1	56.10
Hydrogen	120	0.00

2. Pre-Combustion CO_2 Capture

In this method, the fuel (coal, gas, biomass) is not completely combusted in the reactor, but is converted into a mixture of CO and H_2 in the reforming or gasification process. Subsequently, using the water–gas shift, CO_2 and H_2 are produced. Figure 2 shows a block diagram of the pre-combustion CO_2 capture method in a power plant.

PRE-COMBUSTION

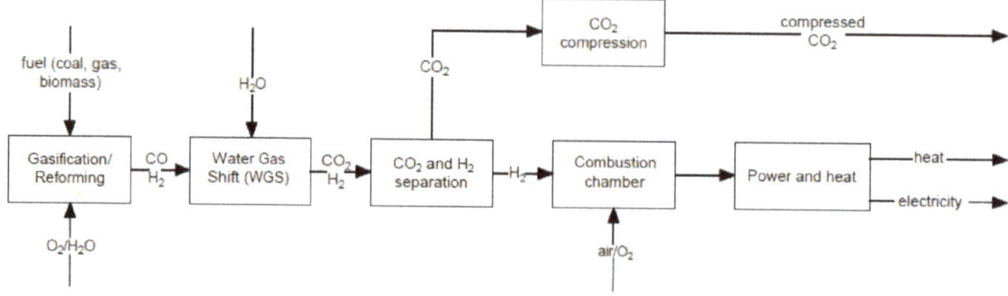

Figure 2. Block diagram of electricity generation heat production with the use of the pre-combustion CO_2 capture method.

Pre-combustion capture is used, e.g., in an integrated gasification combined cycle (IGCC). Carbon dioxide is removed after the gasification process. An example of a typical process for power and heat generation in a gas turbine with pre-combustion capture is shown in Figure 3.

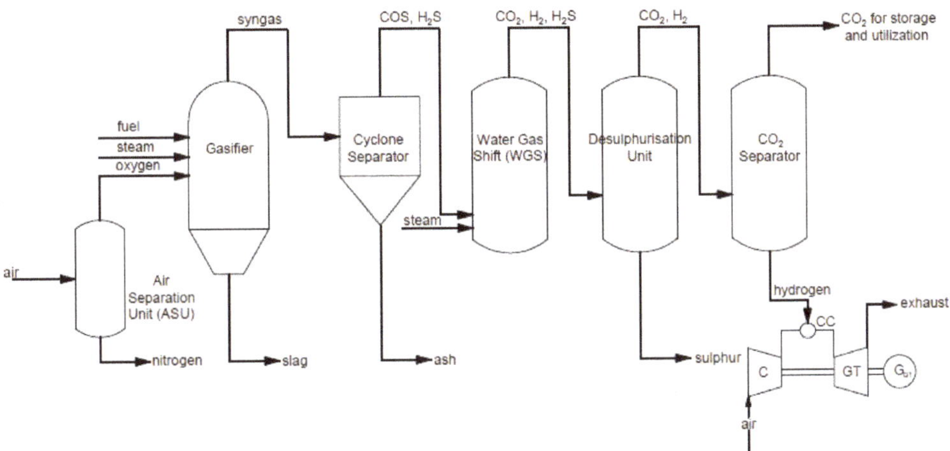

Figure 3. Scheme of integrated gasification combined cycle (IGCC) for electricity generation using a gas turbine using the pre-combustion CO_2 capture method.

In this process, steam and oxygen are provided to the gasifier to produce syngas enriched in hydrogen and carbon monoxide. Then, syngas is sent to a cyclone separator, where it is filtered to remove ash. After, the conversion of syngas and steam to CO_2 and H_2 occurs in the water–gas shift reactor. The received gas needs to be purified of sulfur in the desulfurization unit. Subsequently, CO_2 is captured in the CO_2 separator and is sent for storage or utilization. Received hydrogen is provided to the gas turbine as fuel [18]. Pre-combustion methods are very effective in CO_2 separation on the grounds of the high concentration of CO_2 in fuel before combustion. On the other hand, these processes are expensive due to the need for a gasification unit.

Pre-combustion carbon capture uses physical and chemical methods to capture CO_2 from processed syngas. Chemical absorbents, such as carbonates and physical solvents, such as polypropylene glycol and methanol, are commercially used in industries to capture CO_2. The cost expenditure and energy consumption of carbon capture depend on the utilities and capture process. An effective solvent or absorbent pre-combustion carbon capture technology can achieve more than 90% CO_2 capture, but, at the same time, reduces plant efficiency [19]. The calcium looping process is another method of pre-combustion CO_2 capture, where CO_2 capture is achieved effectively at a low cost. This method involves the sorption of CaO with CO_2 and the desorption of $CaCO_3$ to release CO_2 at an optimal temperature. This cycling process repeats multiple times, and waste heat from the gasifier is used to reduce the heat consumption of the CO_2 capture process. The CaL pre-combustion carbon capture method is highly effective. Low-cost and CO_2 capture are achieved by decreasing energy consumption [20]. The pre-combustion carbon capture demo plant in Port Arthur, United States, has successfully captured 1 million tons of CO_2 since it started operations, without problems. This plant proved that, using the dual pressure swing adsorption (PSA) technology method, purification of hydrogen >99.9% and a high efficiency CO_2 capture can be achieved. When the streaming gas has a low pressure, hydrogen purification is performed and the tail gas is sent to undergo vacuum pressure swing adsorption (VPSA) to separate purified CO_2. If the streaming gas has a high pressure,

CO_2 capture is achieved without VPSA first, and hydrogen purification is achieved from the exiting gas [21].

3. Post-Combustion CO_2 Capture

Post-combustion CO_2 capture methods are based on removing carbon dioxide from flue gas. The capture unit is placed after the purification systems, such as desulphurization, denitrogenation, and dedusting installations. Figure 4 shows a general block diagram of the post-combustion capture technique.

POST-COMBUSTION

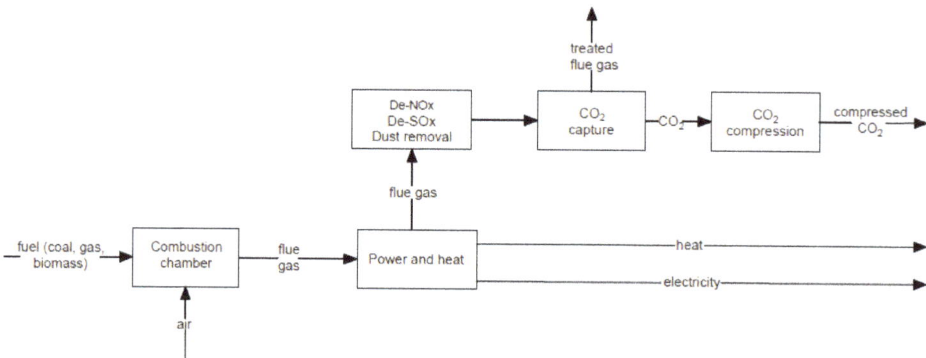

Figure 4. Block diagram of electricity generation and heat production through the use of the post-combustion CO_2 capture method.

In existing conventional power units, post-combustion technologies are the most frequently considered [2,8]; nevertheless, there is one main barrier to using these methods. Since the partial pressure of CO_2 in the flue gas is low (flue gas is under atmospheric pressure and the concentration of carbon dioxide is within 13–15%), the driving force for CO_2 is also low [22]. Post-combustion technologies can be divided up according to the type of process used for capturing carbon dioxide, as follows:

(a) Absorption solvent-based methods

Chemical absorption is the most recognizable method of CO_2 capture. It relies on a reaction between carbon dioxide and a chemical solvent. Solvents that are usually used are alkanolamines, such as monoethanolamine (MEA), diethanolamine (DEA), or methyl diethanolamine (MDEA) in aqueous solution [23]. A schematic diagram of chemical absorption is shown in Figure 5. The process takes place in two stages. In the first stage, the flue gas reacts with the solvent in the absorber to capture CO_2. Subsequently, the rich loading solution is carried to the stripper to regenerate CO_2 at elevated temperatures. The solution without CO_2 (lean-loading solution) is sent back to the absorber column. A high purity carbon dioxide stream from the desorber is transferred for compression and storage or utilization. The chemical absorption process has been used for a long time in the chemical industry. The typically used 30% MEA and MDEA solutions achieve a high process efficiency and a high degree of carbon dioxide purity [23]. The chemical absorption method is a very energy-consuming process due to the need to supply a large amount of heat to the desorber. It is assumed that approx. 30% (37%) of the heat supplied to the steam in the boiler should be directed to the CCS installation in the case of a steam unit fired with hard coal (lignite), depending on the absorber used (for ammonia, the amount of heat needed for regeneration is 22% for hard coal and 27% for lignite). Chemical absorption technologies are used in power plants fired with solid fuel, and they are the only ones

35. Čespiva, J.; Skřínský, J.; Vereš, J. Comparison of potential materials for producer gas wet scrubbing in pilot-scale gasification unit. *WIT Trans. Ecol. Environ.* **2019**, *237*, 87–96.
36. Friedel, P.; Čespiva, J. Reconstructed gasification technology. *Int. J. Mech. Eng.* **2017**, *2*, 129–133.
37. Budner, Z.; Hull, S.; Jodkowski, W.; Sitka, A.; Trawczyński, J.; Walendziewski, J. A proposal of a wasteless method for biomass conversion to electricity Propozycja bezodpadowej metody konwersji biomasy do energii elektrycznej. *Przem. Chem.* **2015**, *1*, 173–176. [CrossRef]
38. Skřínský, J.; Vereš, J.; Čespiva, J.; Ochodek, T.; Borovec, K.; Koloničný, J. Explosion characteristics of syngas from gasification process. *Inz. Miner.* **2020**, *2*, 195–200. [CrossRef]
39. Wnukowski, M. Methods used in tar removal from biomass gasification gas—A review. *Arch. Waste Manag. Environ. Prot.* **2016**, *18*, 17–34.
40. Luo, H.; Lu, Z.; Jensen, P.A.; Glarborg, P.; Lin, W.; Dam-Johansen, K.; Wu, H. Effect of gasification reactions on biomass char conversion under pulverized fuel combustion conditions. *Proc. Combust. Inst.* **2021**, *38*, 3919–3928. [CrossRef]
41. Luo, H.; Lu, Z.; Jensen, P.A.; Glarborg, P.; Lin, W.; Dam-Johansen, K.; Wu, H. Experimental and modelling study on the influence of wood type, density, water content, and temperature on wood devolatilization. *Fuel* **2020**, *260*, 116410. [CrossRef]
42. Kwapinska, M.; Horvat, A.; Xue, G.; Kwapinski, W.; Rabou, L.P.L.M.; Dooley, S.; Leahy, J.J. The Effect of Equivalence Ratio on the Gasification of Torrefied and Un-Torrefied Biomass—Miscanthus x Giganteus. In Proceedings of the European Combustion Meeting 2015, Budapest, Hungary, 30 March–2 April 2015; pp. 1–7.
43. Wnukowski, M.; Jamróz, P. Microwave plasma treatment of simulated biomass syngas: Interactions between the permanent syngas compounds and their influence on the model tar compound conversion. *Fuel Process. Technol.* **2018**, *173*, 229–242. [CrossRef]
44. Jamróz, P.; Kordylewski, W.; Wnukowski, M. Microwave plasma application in decomposition and steam reforming of model tar compounds. *Fuel Process. Technol.* **2018**, *169*, 1–14. [CrossRef]
45. Horvat, A.; Kwapinska, M.; Abdel Karim Aramouni, N.; Leahy, J.J. Solid phase adsorption method for tar sampling—How post sampling treatment affects tar yields and volatile tar compounds? *Fuel* **2021**, *291*, 120059. [CrossRef]
46. Horvat, A.; Kwapinska, M.; Xue, G.; Dooley, S.; Kwapinski, W.; Leahy, J.J. Detailed Measurement Uncertainty Analysis of Solid-Phase Adsorption—Total Gas Chromatography (GC)-Detectable Tar from Biomass Gasification. *Energy Fuels* **2016**, *30*, 2187–2197. [CrossRef]
47. Ciżmiński, P.; Polesek-Karczewska, S.; Kardaś, D. Modeling of drying and pyrolysis in a gasifier during the startup phase. *Trans. Inst. Fluid-Flow Mach.* **2015**, *128*, 73–83.
48. Mularski, J.; Modliński, N. Impact of Chemistry–Turbulence Interaction Modeling Approach on the CFD Simulations of Entrained Flow Coal Gasification. *Energies* **2020**, *13*, 6467. [CrossRef]
49. Li, J.; Burra, K.R.G.; Wang, Z.; Liu, X.; Gupta, A.K. Acid and Alkali Pre-treatment Effects on CO_2 Assisted Gasification of Pine Wood. *J. Energy Resour. Technol.* **2021**, *144*, 022306. [CrossRef]
50. Liu, X.; Burra, K.R.G.; Wang, Z.; Li, J.; Che, D.; Gupta, A.K. Syngas Characteristics From Catalytic Gasification of Polystyrene and Pinewood in CO_2 Atmosphere. *J. Energy Resour. Technol.* **2021**, *143*, 052304. [CrossRef]
51. Sitka, A.; Szulc, P.; Smykowski, D.; Jodkowski, W. Application of poultry manure as an energy resource by its gasification in a prototype rotary counterflow gasifier. *Renew. Energy* **2021**, *175*, 422–429. [CrossRef]
52. Laohalidanond, K.; Kerdsuwan, S.; Burra, K.R.G.; Li, J.; Gupta, A.K. Syngas Generation From Landfills Derived Torrefied Refuse Fuel Using a Downdraft Gasifier. *J. Energy Resour. Technol.* **2021**, *143*, 052102. [CrossRef]
53. Li, J.; Burra, K.G.; Wang, Z.; Liu, X.; Kerdsuwan, S.; Gupta, A.K. Energy Recovery From Composite Acetate Polymer-Biomass Wastes via Pyrolysis and CO_2-Assisted Gasification. *J. Energy Resour. Technol.* **2021**, *143*, 042305. [CrossRef]
54. Sitka, A.; Jodkowski, W.; Szulc, P.; Smykowski, D.; Szumiło, B. Study of the properties and particulate matter content of the gas from the innovative pilot-scale gasification installation with integrated ceramic filter. *Energies* **2021**, *14*, 7476. [CrossRef]
55. Werle, S. Numerical analysis of the combustible properties of sewage sludge gasification gas. *Chem. Eng. Trans.* **2015**, *45*, 1021–1026. [CrossRef]
56. Werle, S.; Dudziak, M. Evaluation of the possibility of the sewage sludge gasification gas use as a fuel. *Ecol. Chem. Eng. S* **2016**, *23*, 229–236. [CrossRef]
57. Szwaja, S.; Kovacs, V.B.; Bereczky, A.; Penninger, A. Sewage sludge producer gas enriched with methane as a fuel to a spark ignited engine. *Fuel Process. Technol.* **2013**, *110*, 160–166. [CrossRef]
58. Werle, S.; Dudziak, M. Analysis of organic and inorganic contaminants in dried sewage sludge and by-products of dried sewage sludge gasification. *Energies* **2014**, *7*, 462–476. [CrossRef]
59. Pawlak-Kruczek, H.; Wnukowski, M.; Niedzwiecki, L.; Czerep, M.; Kowal, M.; Krochmalny, K.; Zgora, J.; Ostrycharczyk, M.; Baranowski, M.; Tic, W.J.; et al. Torrefaction as a Valorization Method Used Prior to the Gasification of Sewage Sludge. *Energies* **2019**, *12*, 175. [CrossRef]
60. Ziółkowski, P.; Madejski, P.; Amiri, M.; Kuś, T.; Stasiak, K.; Subramanian, N.; Pawlak-Kruczek, H.; Badur, J.; Niedźwiecki, Ł.; Mikielewicz, D. Thermodynamic Analysis of Negative CO_2 Emission Power Plant Using Aspen Plus, Aspen Hysys, and Ebsilon Software. *Energies* **2021**, *14*, 6304. [CrossRef]
61. Mountouris, A.; Voutsas, E.; Tassios, D. Plasma gasification of sewage sludge: Process development and energy optimization. *Energy Convers. Manag.* **2008**, *49*, 2264–2271. [CrossRef]

62. Kordylewski, W.; Michalski, J.; Ociepa, M.; Wnukowski, M. A microwave plasma potential in producer gas cleaning—Preliminary results with a gas derrived from a sewage sludge. In Proceedings of the VI Konferencja Naukowo-Techniczna Energetyka Gazowa 2016, Zawierciu, Poland, 20–22 April 2016; Silesian University of Technology: Gliwice, Poland, 2016.

63. Striūgas, N.; Valinčius, V.; Pedišius, N.; Poškas, R.; Zakarauskas, K. Investigation of sewage sludge treatment using air plasma assisted gasification. *Waste Manag.* **2017**, *64*, 149–160. [CrossRef]

64. Brachi, P.; Di Fraia, S.; Massarotti, N.; Vanoli, L. Combined heat and power production based on sewage sludge gasification: An energy-efficient solution for wastewater treatment plants. *Energy Convers. Manag. X* **2022**, *13*, 100171. [CrossRef]

65. Blasenbauer, D.; Huber, F.; Lederer, J.; Quina, M.J.; Blanc-Biscarat, D.; Bogush, A.; Bontempi, E.; Blondeau, J.; Chimenos, J.M.; Dahlbo, H.; et al. Legal situation and current practice of waste incineration bottom ash utilisation in Europe. *Waste Manag.* **2020**, *102*, 868–883. [CrossRef]

66. Ziółkowski, P.; Badur, J.; Pawlak- Kruczek, H.; Stasiak, K.; Amiri, M.; Niedzwiecki, L.; Krochmalny, K.; Mularski, J.; Madejski, P.; Mikielewicz, D. Mathematical modelling of gasification process of sewage sludge in reactor of negative CO_2 emission power plant. *Energy* **2022**, *244*, 122601. [CrossRef]

67. Werle, S. Sewage sludge gasification: Theoretical and experimental investigation. *Environ. Prot. Eng.* **2013**, *39*, 25–32. [CrossRef]

68. Werle, S. Impact of feedstock properties and operating conditions on sewage sludge gasification in a fixed bed gasifier. *Waste Manag. Res.* **2014**, *32*, 954–960. [CrossRef] [PubMed]

69. Schweitzer, D.; Gredinger, A.; Schmid, M.; Waizmann, G.; Beirow, M.; Spörl, R.; Scheffknecht, G. Steam gasification of wood pellets, sewage sludge and manure: Gasification performance and concentration of impurities. *Biomass Bioenergy* **2018**, *111*, 308–319. [CrossRef]

70. Akkache, S.; Hernández, A.B.; Teixeira, G.; Gelix, F.; Roche, N.; Ferrasse, J.H. Co-gasification of wastewater sludge and different feedstock: Feasibility study. *Biomass Bioenergy* **2016**, *89*, 201–209. [CrossRef]

71. Wnukowski, M.; Jamróz, P.; Niedzwiecki, L. The role of hydrogen in microwave plasma valorization of producer gas. *Int. J. Hydrogen Energy* **2021**, *in press.* [CrossRef]

72. Chun, Y.N.; Song, H.G. Microwave-enhanced gasification of sewage sludge waste. *Environ. Eng. Res.* **2018**, *24*, 591–599. [CrossRef]

Review

Methods and Techniques for CO_2 Capture: Review of Potential Solutions and Applications in Modern Energy Technologies

Paweł Madejski *, Karolina Chmiel, Navaneethan Subramanian and Tomasz Kuś

Department of Power Systems and Environmental Protection Facilities, Faculty of Mechanical Engineering and Robotics, AGH University of Science and Technology, 30-059 Kraków, Poland; kachmiel@agh.edu.pl (K.C.); subraman@agh.edu.pl (N.S.); kus@agh.edu.pl (T.K.)
* Correspondence: pawel.madejski@agh.edu.pl

Abstract: The paper presents and discusses modern methods and technologies of CO_2 capture (pre-combustion capture, post-combustion capture, and oxy-combustion capture) along with the principles of these methods and examples of existing and operating installations. The primary differences of the selected methods and technologies, with the possibility to apply them in new low-emission energy technologies, were presented. The following CO_2 capture methods: pre-combustion, post-combustion based on chemical absorption, physical separation, membrane separation, chemical looping combustion, calcium looping process, and oxy-combustion are discussed in the paper. Large-scale carbon capture utilization and storage (CCUS) facilities operating and under development are summarized. In 2021, 27 commercial CCUS facilities are currently under operation with a capture capacity of up to 40 Mt of CO_2 per year. If all projects are launched, the global CO_2 capture potential can be more than ca. 130–150 Mt/year of captured CO_2. The most popular and developed indicators for comparing and assessing CO_2 emission, capture, avoiding, and cost connected with avoiding CO_2 emissions are also presented and described in the paper.

Keywords: carbon capture and storage installation; CO_2 capture methods; CO_2 emission level assessment indicators

Citation: Madejski, P.; Chmiel, K.; Subramanian, N.; Kuś, T. Methods and Techniques for CO_2 Capture: Review of Potential Solutions and Applications in Modern Energy Technologies. *Energies* **2022**, *15*, 887. https://doi.org/10.3390/en15030887

Academic Editor: João Fernando Pereira Gomes

Received: 2 December 2021
Accepted: 20 January 2022
Published: 26 January 2022

Publisher's Note: MDPI stays neutral with regard to jurisdictional claims in published maps and institutional affiliations.

1. Introduction

With the increase in electricity consumption around the world, electricity demands are increasing every day. During electricity generation using energy technologies based on fossil fuels, the emission of harmful pollutants into the environment (gaseous, liquid, and solid) occurs as the emission of NOx, SOx, dust, CO_2, and wastewater (e.g., from flue-gas treatment installations) [1–3]. Last year, a great deal of effort in modern low-emission energy technologies was directed at activities leading to decreased gaseous pollutant emissions [4,5]. The emission of carbon dioxide (CO_2), which is treated as one of the main reasons for global warming when fossil fuel is burned, cannot be avoided. Fossil-fueled power-production technology plays a significant role in contributing to the emission of greenhouse gases into the atmosphere. By reducing the emission of CO_2 into the atmosphere, and by switching to an alternative power generation with zero-emission, it is possible to prevent future catastrophic effects. The carbon capture utilization and storage (CCUS) methods and technologies are among the many ways to reduce CO_2 emissions. CCUS technologies aim to capture CO_2 from large industrial sources and store it in underground structures, or use it through conversion into useful products [6]. All this is happening while emissions from the industrial and energy sectors are reduced, which makes this process one of the most current scientific research endeavors, while also presenting socio-economic challenges. The current fees related to CO_2 emissions in the European Union amount to over 50 EUR/tCO_2 [7], and there is an expected upward trend for coming years, forced by political declarations and treaties.

There are four different ways to reduce CO_2 emission levels [8]:

1. Reducing the use of fossil fuels by:
    - improving the efficiency of energy conversion processes;
    - reducing the demand for energy;
    - using renewable (non-fossil fuel) energy sources, such as hydropower, wind, biomass, solar cells, and nuclear power;
    - increasing the use of green hydrogen, which is produced by splitting water using electricity from renewable energy.
2. Replace technologies using fossil fuels with a low carbon to hydrogen C/H_2 ratio by replacing coal and oil with gaseous fuels.
3. Capturing CO_2 from fuel combustion in power plants and other industrial processes and storing it in appropriate geological structures, in exhausted or exploited gas or crude oil deposits (intensification of crude oil extraction, enhanced oil recovery (EOR)), or at the bottom of oceans.
4. Limiting deforestation processes and thus storing more CO_2 in biomass.

Carbon capture utilization and storage (CCUS) is a family of methods to reduce the emission of CO_2 from fossil-fueled power plants. The CCS can be coupled with Power Plants (PPs) and Combined Heat and Power Plants (CHPPs) to reduce the emission of CO2 in the flue gas. First-generation carbon capture technology had a lower efficiency in carbon capture, and was challenging to integrate with the complex structures of a power plant. With improved research and development, the second- and third-generation power plants using carbon capture technology showed improved efficiency and a low cost compared to first-generation CCS technology [9]. The three different methods (pre-, post-, and oxyfuel combustion) for CO_2 capture and separation are under development. The oxyfuel combustion method is considered a promising solution from an energy-efficiency power-generation point of view. Authors have noted, that the energy penalty for the oxy-combustion method can be around 4%, in comparison to 8–12% for post-combustion methods [10].

The implementation of CCS in Europe is focused on two major factors, the development of power generation technology with carbon capture at a low cost and selling CO_2 at a high price to reimburse the cost of CO_2 transportation. In 2008, the European Parliament approved the "draft CCS directive", which aims to guide CO_2 geological storage. Due to public opposition from European Union countries for underground CO_2 storage, many countries allow only offshore storage projects [11]. CCUS technologies are considered as crucial technologies for the European Commission, and are explicitly included in, e.g., the European Green Deal. Nowadays, more and more important projects at the industrial scale are funded by the Innovation Fund (https://ec.europa.eu/clima/eu-action/funding-climate-action/innovation-fund/large-scale-projects_en, accessed on 20 January 2022). The International Energy Agency forecasts that CCS will contribute up to 21% of the reduction in CO_2 emissions into the atmosphere. Many countries, including Asian countries, still depend on coal-fired power production because of the low cost and reliability, and these cannot be wholly replaced with renewable energy systems. Some European countries, such as Poland, will depend on fossil-fuel power production for at least 30 more years. According to Polish Energy Policy, by the year 2050, electricity demand in Poland will be produced by renewable energy and future nuclear power projects. The use of coal for the next few years makes CCS technology inevitable [12]. Carbon capture and storage is a method for capturing the concentrated CO_2 in flue gas from fossil-fueled power plants and store them in one place. There are three methods of CO_2 capture: pre-combustion carbon capture, post-combustion carbon capture, and the oxy-combustion carbon capture method.

- Pre-combustion carbon capture occurs before the combustion process (through fuel gasification with oxygen, e.g., integrated IGCC coal gasification technology).
- Post-combustion carbon capture occurs after the combustion process (capturing CO_2 from flue gas, e.g., using chemical absorption, physical adsorption, membrane separation, or the use of a chemical loop).

- Oxy-combustion carbon capture occurs after the combustion process in an oxygen atmosphere by separating CO2 generated during the oxy-combustion process, e.g., using an oxygen gas turbine. Oxygen atmosphere can be obtained by removing nitrogen from the air before the combustion process.

A diagram explaining the methods and techniques for CO_2 capture is shown in Figure 1.

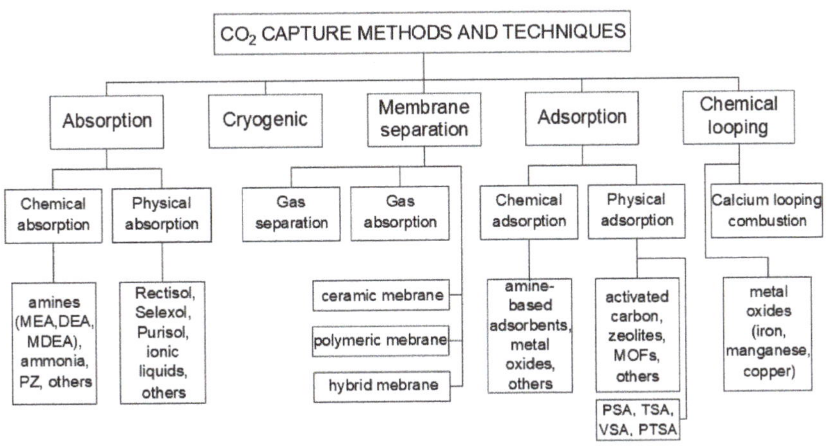

Figure 1. Methods and techniques of CO_2 capture.

In pre-combustion carbon capture, fuel is oxidized using a gasification process, which produces syngas with a composition of hydrogen (H_2) and carbon monoxide (CO). The produced CO is converted into CO_2, which is captured before combustion. Shijaz et al. showed a comparison between power generation using coal gasification without carbon capture, pre-combustion carbon capture, or chemical looping combat (CLC). The results showed that the overall efficiency of a power plant with pre-combustion carbon capture and CLC is reduced compared to a plant without CO_2 capture. CO_2 captured from fuel reduces the fuel volume sent to the turbine, reducing power generation. However, due to environmental concerns, including a CO_2 capture unit is unavoidable [13]. Mukherjee et al. compared each type of carbon capture method using an IGCC power plant without CO_2 capture. CLC was combined with a coal-fired IGCC, and analyses were performed based on electrical efficiency and carbon capture efficiency. The CLC and oxyfuel combustion methods showed a value of around a 100% carbon capture rate compared to pre-combustion CO_2 capture, which achieved 94.8% CO_2 capture. Another method of coal direct chemical looping combustion, where coal is fed directly to a boiler without gasification, increased the electrical efficiency and achieved 100% CO_2 capture. The results from the comparison of IGCC-CLC, pre-combustion, and oxyfuel combustion showed that the methods' energy penalties were 4.5%, 7.1%, and 9.1%, respectively [14]. This article describes the various carbon capture and storage methods and technology used in large-scale units. The CO_2 emission from a coal-fired ultra-supercritical power plant is calculated.

The SO_2 and NO_x content in the flue gas has a higher chance of affecting the purity of CO_2 during CO_2 capture. The high purity of the CO_2 stream is very important for recycling methods. A pilot, dual-reflux VPSA unit, installed in the Łagisza Power plant in Poland, was installed for post-combustion CO_2 capture. Before the flue gas from the boiler is sent to the DR-VPSA unit, the flue gas is passed through an absorber, an adsorber, and a glycolic gas dehydration system to remove SO_2 and NO_x. Activated carbon works effectively for the removal of SO_2 compared to the removal of NO_x, which leads to a high-purity CO_2 [15].

CO_2 emissions from the power sector are mainly caused by modern technologies, such as coal-fired, gas-fired, oil-fired, and combined cycle gas turbine (CCGT) power plants. Table 1 presents the CO_2 emissions and lower heating value depending on fossil fuel. From Table 1, it can be seen that CO_2 emissions depend on the content of the fuel. The higher the fuel content, the higher the CO_2 emissions will be. The increasing share of H_2 gives better properties to fossil fuels, taking into account the LHV and CO_2 emission levels. Natural gas consists mainly of CH_4 and is characterized by almost two times lower emissions than hard and lignite coal, and an almost two times larger LHV. This fact comes from gas fuel composition, where four atoms of hydrogen are inside the methane molecule of every carbon atom. Other modern power generating technologies, such as nuclear, renewable energy sources (RES), and hydrogen-based technologies, are less likely to produce emissions. A newly built CCGT power plant has CO_2 emissions of 350 kg_{CO2}/MWh without carbon capture, indicating the same emissions as a gas-fired power plant [16]. In the case of an ultra-supercritical power plant fired with coal, CO_2 emissions up to 700 $kgCO_2$/MWh can be produced. According to the type of coal-fired critical power plant used, CO_2 emissions range from 690 to 830 $kgCO_2$/MWh [17]. The use of fossil fuel in modern energy technology will continue until it is replaced with alternative technologies, such as renewable energies and power production without emission. Until these technologies are replaced with those without CO_2 emissions, CO_2 capture is unavoidable to reduce greenhouse gases and protect the environment.

Table 1. Emission of carbon dioxide (CO_2) during combustion of different hydrocarbon fuels [1–3,8].

Fuel	LHV	Emission
	MJ/kg	kgCO₂/GJ
Hard coal	>23.9	94.60
Lignite	<17.4	101.20
Crude oil	43.0	74.07
Petrol	43.4	66.00
Paraffin oil	41.5	71.50
Heating oil	42.8	77.37
Diesel	42.6	74.07
Natural gas	47.1	56.10
Hydrogen	120	0.00

2. Pre-Combustion CO_2 Capture

In this method, the fuel (coal, gas, biomass) is not completely combusted in the reactor, but is converted into a mixture of CO and H_2 in the reforming or gasification process. Subsequently, using the water–gas shift, CO_2 and H_2 are produced. Figure 2 shows a block diagram of the pre-combustion CO_2 capture method in a power plant.

PRE-COMBUSTION

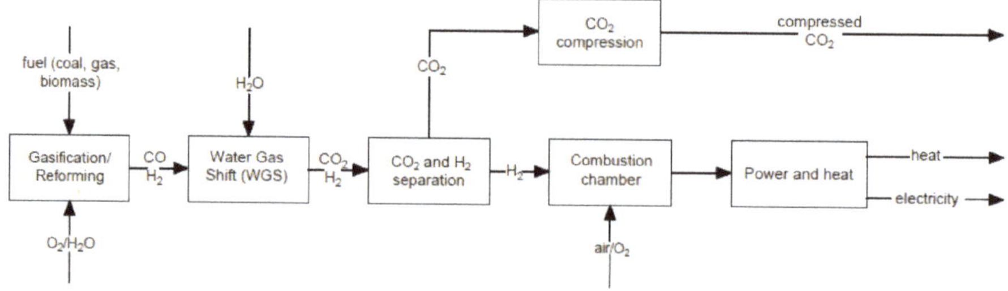

Figure 2. Block diagram of electricity generation heat production with the use of the pre-combustion CO_2 capture method.

Pre-combustion capture is used, e.g., in an integrated gasification combined cycle (IGCC). Carbon dioxide is removed after the gasification process. An example of a typical process for power and heat generation in a gas turbine with pre-combustion capture is shown in Figure 3.

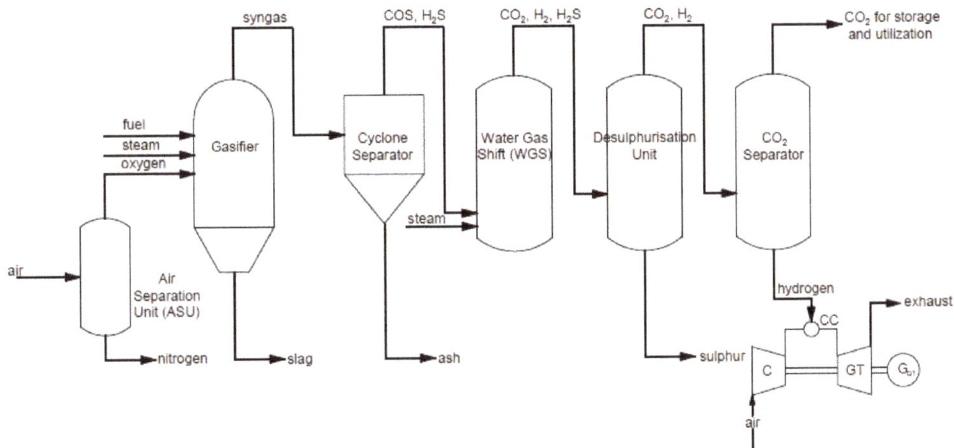

Figure 3. Scheme of integrated gasification combined cycle (IGCC) for electricity generation using a gas turbine using the pre-combustion CO_2 capture method.

In this process, steam and oxygen are provided to the gasifier to produce syngas enriched in hydrogen and carbon monoxide. Then, syngas is sent to a cyclone separator, where it is filtered to remove ash. After, the conversion of syngas and steam to CO_2 and H_2 occurs in the water–gas shift reactor. The received gas needs to be purified of sulfur in the desulfurization unit. Subsequently, CO_2 is captured in the CO_2 separator and is sent for storage or utilization. Received hydrogen is provided to the gas turbine as fuel [18]. Pre-combustion methods are very effective in CO_2 separation on the grounds of the high concentration of CO_2 in fuel before combustion. On the other hand, these processes are expensive due to the need for a gasification unit.

Pre-combustion carbon capture uses physical and chemical methods to capture CO_2 from processed syngas. Chemical absorbents, such as carbonates and physical solvents, such as polypropylene glycol and methanol, are commercially used in industries to capture CO_2. The cost expenditure and energy consumption of carbon capture depend on the utilities and capture process. An effective solvent or absorbent pre-combustion carbon capture technology can achieve more than 90% CO_2 capture, but, at the same time, reduces plant efficiency [19]. The calcium looping process is another method of pre-combustion CO_2 capture, where CO_2 capture is achieved effectively at a low cost. This method involves the sorption of CaO with CO_2 and the desorption of $CaCO_3$ to release CO_2 at an optimal temperature. This cycling process repeats multiple times, and waste heat from the gasifier is used to reduce the heat consumption of the CO_2 capture process. The CaL pre-combustion carbon capture method is highly effective. Low-cost and CO_2 capture are achieved by decreasing energy consumption [20]. The pre-combustion carbon capture demo plant in Port Arthur, United States, has successfully captured 1 million tons of CO_2 since it started operations, without problems. This plant proved that, using the dual pressure swing adsorption (PSA) technology method, purification of hydrogen >99.9% and a high efficiency CO_2 capture can be achieved. When the streaming gas has a low pressure, hydrogen purification is performed and the tail gas is sent to undergo vacuum pressure swing adsorption (VPSA) to separate purified CO_2. If the streaming gas has a high pressure,

CO_2 capture is achieved without VPSA first, and hydrogen purification is achieved from the exiting gas [21].

3. Post-Combustion CO_2 Capture

Post-combustion CO_2 capture methods are based on removing carbon dioxide from flue gas. The capture unit is placed after the purification systems, such as desulphurization, denitrogenation, and dedusting installations. Figure 4 shows a general block diagram of the post-combustion capture technique.

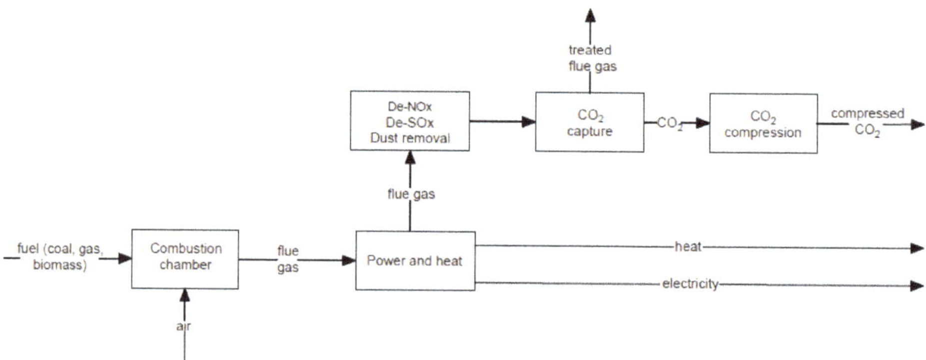

Figure 4. Block diagram of electricity generation and heat production through the use of the post-combustion CO_2 capture method.

In existing conventional power units, post-combustion technologies are the most frequently considered [2,8]; nevertheless, there is one main barrier to using these methods. Since the partial pressure of CO_2 in the flue gas is low (flue gas is under atmospheric pressure and the concentration of carbon dioxide is within 13–15%), the driving force for CO_2 is also low [22]. Post-combustion technologies can be divided up according to the type of process used for capturing carbon dioxide, as follows:

(a)   Absorption solvent-based methods

Chemical absorption is the most recognizable method of CO_2 capture. It relies on a reaction between carbon dioxide and a chemical solvent. Solvents that are usually used are alkanolamines, such as monoethanolamine (MEA), diethanolamine (DEA), or methyl diethanolamine (MDEA) in aqueous solution [23]. A schematic diagram of chemical absorption is shown in Figure 5. The process takes place in two stages. In the first stage, the flue gas reacts with the solvent in the absorber to capture CO_2. Subsequently, the rich loading solution is carried to the stripper to regenerate CO_2 at elevated temperatures. The solution without CO_2 (lean-loading solution) is sent back to the absorber column. A high purity carbon dioxide stream from the desorber is transferred for compression and storage or utilization. The chemical absorption process has been used for a long time in the chemical industry. The typically used 30% MEA and MDEA solutions achieve a high process efficiency and a high degree of carbon dioxide purity [23]. The chemical absorption method is a very energy-consuming process due to the need to supply a large amount of heat to the desorber. It is assumed that approx. 30% (37%) of the heat supplied to the steam in the boiler should be directed to the CCS installation in the case of a steam unit fired with hard coal (lignite), depending on the absorber used (for ammonia, the amount of heat needed for regeneration is 22% for hard coal and 27% for lignite). Chemical absorption technologies are used in power plants fired with solid fuel, and they are the only ones that are commercially available. It is assessed that the amine method can capture approx. 85–95% carbon dioxide included in flue gas with a purity above 99.95% [2].

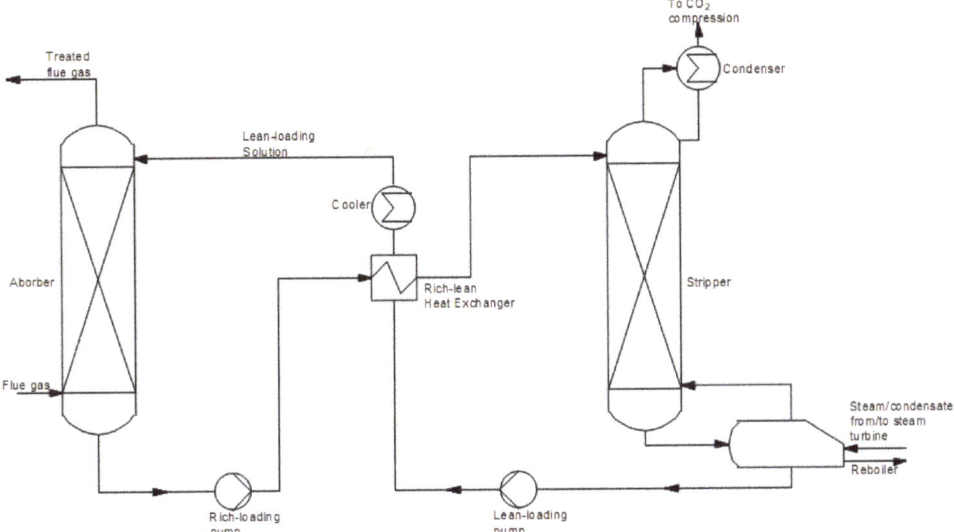

Figure 5. Scheme of the post-combustion CO_2 capture method using a chemical absorption process (based on [22]).

Nowadays, apart from conventional solvents (amine-based-MEA, DEA, ammonia, piperazine), there are other solvents developed for the CO_2 capture process. Solvent blends offer the ability to improve absorption properties by combining types properly. Primary and secondary amines have high absorption rates, and tertiary amines are characterized by a high capacity [8]. For example, blending MEA with a little PZ can improve the absorption rate (PZ is 50 times faster than MEA) [24]. Another possibility is to use a solution of 2-amino-2-methyl1-propanol (AMP) promoted with PZ. Artanto et al. showed that a mixture of 25 wt% AMP and 5 wt% PZ can be a good substitution for MEA [25]. Ionic liquids (ILs) are novel alternatives for amines. These low melting salts are comprised of a large organic cation and an arbitrary anion, which can be combined freely, obtaining a great variety of compound properties. ILs can physically or chemically absorb CO_2, depending on pressure [8]. A review [26] and articles [24,27] have presented deep insight into this technology. In the case of reducing energy consumption, there are new generation solvents—water-free solvents and biphasic solvents—that have been proposed. The presence of water in a solvent enhances the energy demand for the regeneration process. Novel water-free solvents, such as non-aqueous organic amine blends (methanol, ethylene glycol), aminosilicones, or amines with a superbase have been observed [24]. In [28], researchers showed that solvent mixtures based on ethylene glycol, used in the chemisorption process, can achieve CO_2 capture efficiencies of up to 95%. Deep eutectic solvents (DESs), such as choline chloride and ethylene glycol at a 1:2 mole ratio are getting more attention. They are fluids consisting of organic halide salts and metal salts or a hydrogen bond donor. DESs have similar properties to ILs, but they are cheaper and environmentally friendlier. According to [29], using DESs can decrease the vapor pressure of a solvent, achieve a lower effect of corrosion, and needs less energy in the regeneration process.

The physical absorption method is based on using a chemically inert solvent, which absorbs CO_2 physically. Absorption occurs in water or organic absorbers (methanol, N-methyl-2-pyrrolidone, dimethyl ether). This method achieves the best results for low temperatures and high pressures of the separated gas. Therefore, it is used to capture CO_2 from the coal gasification process. In this method, there are distinguished processes, with the use of solvent such as Selexol™, Rectisol™, Ifpexol™, Fluor™, Purisol™, Sulfinol™, and Morphysorb™ [23,24,30].

(b)   Adsorption–physical separation

Adsorption is a process that uses a solid surface to remove carbon dioxide from a mixture. Physical separation relies on adsorption, absorption, and cryogenic separation methods. It can be physical (Van der Waals forces for adhesion CO_2—physisorption) or chemical (covalent bonding between compounds—chemisorption) [31]. Physical adsorption uses various porous materials (such as activated carbon, alumina, metallic oxides, or zeolites [6]) to absorb carbon dioxide. Activated carbons contain amorphous carbon, and it is low-cost material with the advantage of having a large surface area and the possibility of modifying its pore structure. However, the weak binding energy with carbon dioxide causes this material to need to be highly microporous to be useful for carbon capture [8]. Zeolites (crystalline aluminosilicates) have good adsorption properties for CO_2 capture, but they are hydrophilic. The presence of water weakens these properties by reducing the strength of interactions between coupled compounds [26]. A new approach is to use metal–organic frameworks (MOFs) in adsorption processes. MOFs consist of metal ions or ion clusters linked by organic ligands and bridges that create strong coordination bonds. On account of this, MOFs are characterized as having benefits such as ease of design and synthesis, a high porosity, and tailored pore properties [32]. One of the other adsorption materials is silica. Silicas are non-carbonaceous substances with a large surface area and pore size, and they are highly mechanically stable. Mesoporous silica materials use amine-based substances for CO_2 capture [8,31]. The methods of adsorption are as follows: pressure swing adsorption (PSA), temperature swing adsorption (TSA), vacuum swing adsorption (VSA), and pressure–temperature swing adsorption (PTSA).

(c)   Membrane separation

Figure 6 depicts the membrane separation process. In the first place, flue gas is directed to an absorber to cool to the operating temperature of the membrane. Subsequently, flue gas is transported to the membrane. This method uses a spiral wound, flat sheet, and hollow fiber modules [30].

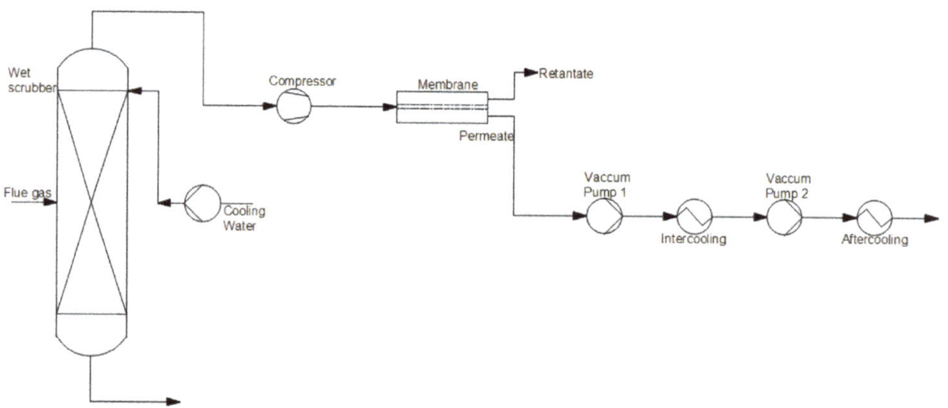

Figure 6. Scheme of the post-combustion CO_2 capture method using a membrane separation process (based on [22]).

There are two types of membrane capture technology: gas separation membranes and gas absorption membranes. With a gas separation membrane, gas with CO_2 is introduced at the high-pressure side of the membrane. Carbon dioxide is recovered at the low-pressure side. A solid microporous membrane is used to enable gas flow and absorption in the gas absorption system. This system has a high removal rate of CO_2, on the grounds of minimization of flooding, foaming, channeling, and entrainment. The principles of both membrane systems are shown in Figure 7 [22,30].

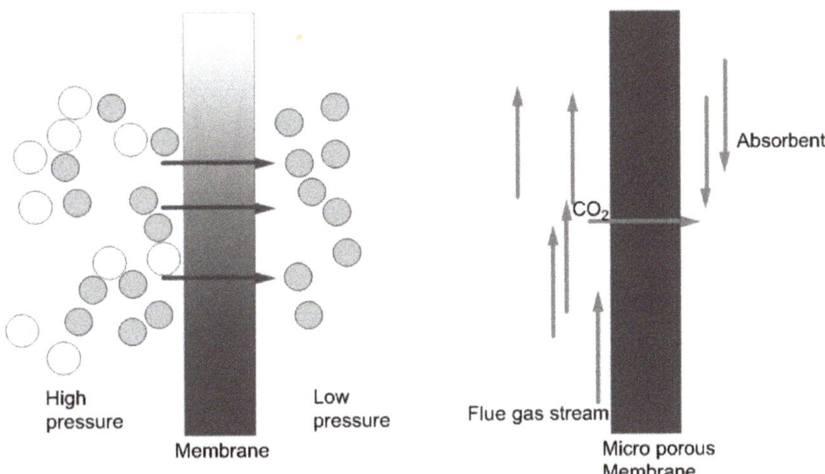

Figure 7. Principle of (**left**) a gas separation membrane, (**right**) a gas absorption membrane (based on [30]).

Membranes should characterize relevant properties for gas separation—proper permeability and selectivity. There are three types of membrane materials: polymeric membranes (organic), ceramic membranes (inorganic), and hybrid membranes [22]. A polymeric membrane has a lower cost of production than the others with a relatively high gas flux and it is mechanically stable [33]. Nevertheless, it has generally low selectivity CO_2/N_2—less than 100, and it is supposed to be 200 [32]. Ceramic membranes, especially zeolites and their derivatives, obtain high selectivities. However, the production of ceramic membranes is more difficult [22]. Hybrid membranes (modified on the surface of inorganic membranes) provide advantages of both membranes, polymeric and ceramic. They have the flexibility and low cost of production of a polymer and the high selectivity of an inorganic material [8,22]. For post-combustion capture, commercially available polymeric membranes, such as PRISM, Polaris, PolyActive, PermSelect, and Medal, are introduced in [34]. A new approach is to use metal–organic frameworks (MOFs) in the experimental stage. These offer many properties that are useful for membranes, such as large surface areas, adjustable pore sizes, and controllable pore-surface properties [32].

(d)   Chemical looping combustion (CLC) and calcium looping process (CLP)

Chemical looping technology uses two reactors, an air reactor and a fuel reactor. These reactors typically circulate fluidized beds that are coupled for carrier transport. In the air reactor, oxidation of an "oxygen carrier", usually metal particles, such as iron, manganese, or copper, occurs with the oxygen from air. As a result of the reaction, metal oxides are formed. These compounds are carried to a second reactor, where they react with the fuel. Metal oxides are reduced during combustion, producing energy and flue gas as a stream of CO_2 and H_2O. The flue gas can be condensed to receive pure CO_2 [35,36].

The calcium looping process is a type of chemical looping. The process (Figure 8) is based on a reversible reaction between calcium oxide and carbon dioxide. The reaction of bounding CaO and CO_2 is called carbonation, and takes place in the first reactor. Subsequently, the formed calcium carbonate in the carbonator is transported to the second reactor, called a calciner, where the reversible reaction occurs and high purity CO_2 stream is produced (>95%). In the calciner, the heat needed for the reaction is generated by burning fuel in the oxygen atmosphere, and sometimes the CLC capture method is considered a kind of oxy-combustion method. The reactors (circulating fluidized bed (CFB)) are coupled to transport solid and cyclones separate solid and gaseous mass streams. Calcium looping technology has a few advantages. It uses a cheap sorbent (lime) and the flue gas is partially desulfurized. Moreover, the process uses fluidized beds, and this mature high-temperature technology can generate power [6,37].

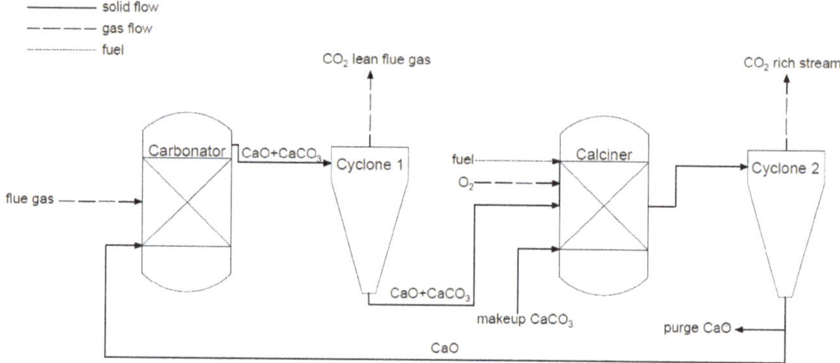

Figure 8. Scheme of the calcium looping process (adapted from [37]).

(e)   Cryogenic method

The cryogenic method of carbon capture technology uses liquefied natural gas (LNG) to provide cold energy to capture CO_2. Cryogenic CO_2 capture is used in oxyfuel combustion technology as well as in post-combustion carbon capture technology to separate CO_2 from flue gas. With cryogenic CO_2 capture, it is possible to produce high purity CO_2 of up to 99.17%. This method includes a few processes, such as compression, expansion, separation, and cooling. The cryogenic method is less preferred because of its high operational cost [38]. The cryogenic method used in post-combustion carbon capture is carried out using various methods [38,39].

(f)   Application of absorption-based post-combustion capture method

The absorption-based post-combustion capture is the most widely used method due to its efficiency and lower energy consumption. Monoethanolamine (MEA), methyl diethanol amine (MDEA) and piperazine (PZ) are the most extensively used amine solvents in large-scale industries [40]. There are many classified technologies used in adsorption and absorption, as well as in membrane separation. Chao et al. analyzed the challenges and compared the commercial use of PCC technology using solvents for absorbents, bed configurations for adsorption, and membrane processes. Among the adsorbent processes, temperature swing adsorption is very effective for both adsorptions, using solid and solvents, compared to pressure swing adsorption (PSA) and vacuum swing adsorption (VSA). TSA is more efficient than PSA and VSA, but it consumes a large amount of energy during regeneration. In the case of chemical absorption, MEA is the best and most used solvent for CO_2 capture. MEA shows a good absorption and desorption rate when mixed with other solvents as well. However, the solvent absorption process requires a high energy consumption for the regeneration of solvents, and solvent losses may occur due to evaporation and chemical degradation, leading to reduced absorption capacity [41]. Lungkadee et al. showed simulation analyses of retrofitting a post-carbon capture unit with a 300 MW power plant. The amine-based PCC technology used MEA amine for the carbon capture process, and was estimated to cost less than 55 $/ton of CO_2 capture. The absorber and desorber used in this process were designed to have 90% CO_2 capture capacity and 30 wt% of MEA. About 63.075 kg/s of CO_2 was captured from flue gas at a flow rate of 458 kg/s with 15% CO_2 content [42]. Another simulation analysis of natural gas combined cycle (NGCC) power plants with PZ solvent showed better performance when compared to that of MEA solvent. The use of 40 wt% PZ solvent showed significant improvements in capture efficiency, energy consumption, and capture cost compared to that of using 30 wt% MEA solvent. The lowest CO_2 capture cost was obtained at 34.65 \$/ton of CO_2 using 40 wt% PZ solvent [43]. Hadri et al. showed a comparison of 30 different amine solutions (30 wt%) used for post-combustion carbon capture. The amine solutions were analyzed using a solvent screening setup, where amine was passed at 1 bar pressure with a gas containing 15% CO_2, and CO_2 loading was measured. Compared to that of other amines, hexamethylenediamine had the highest CO_2 loading of 1.35 and triethanolamine had the lowest CO_2 loading of 0.39 [44].

(g)   Converting CO_2 into value-added chemicals

CO_2 can be utilized to satisfy the needs of various industries as fuels and chemicals or beverages and food [45]. Technologies that allow to convert CO_2 into value-added chemicals are still being developed because of their economic and environmental benefits. In contrast to physical processes, the valence state of CO_2 changes [46]. This process can be used to produce chemical feedstock (polymers, plastics, carbonates [47]), as well as energy carriers (methane, ethane, methanol, syngas). Among the chemical conversions, it is possible to distinguish thermochemical, electrochemical (photoelectrochemical [48]), and biological processes, where enzymes are used [49]. Because of the high stability of CO_2, there is a thermodynamic barrier in the CO_2 conversion process [50]. A crucial component in most processes connected with converting CO_2 into value-added chemicals is hydrogen. It should be produced using renewable energy sources to maintain an environmentally friendly effect.

4. Oxy-Combustion CO_2 Capture

The exhaust gas from combustion in an oxygen-enriched atmosphere (oxy-combustion) consists mainly of carbon dioxide and water vapor (nitrogen content is minimized). From the condensation of water vapor, CO_2 separation is possible. The condensation temperature is higher than ambient conditions, except for very low partial pressures during the condensation process. The oxygen for combustion is produced using the air separation process, which gives an oxygen purity of about 95%. The general scheme for the process using the oxy-combustion method is presented in Figure 9.

Application of oxy-combustion technology mainly concerns solid fuel-fired boilers, including pulverized coal boilers (PCs) or circulating fluidized bed boilers (CFBs), but more and more consideration is being given to the possibility of using oxy-combustion in energy systems with gas turbines. In oxy-combustion technologies, low-temperature and high-temperature boilers can be distinguished. The combustion process in low-temperature boilers usually takes place in oxygen mixed with recirculated exhaust gases. The flame temperatures are similar to those of air-powered units. In the case of high-temperature boilers, the temperature can exceed 2400 °C.

The strengths of oxy-combustion are nitrogen oxide (NO_x) reduction, boiler dimension reductions, a simplified CO_2 capture method compared to other technologies, the possibility of applying in existing technologies, and less mass flow rate of exhaust gases (about 75% less compared to combustion in air). The weaknesses of oxy-combustion are the high material requirements because of the high temperatures, an efficiency decrease (oxygen production process is energy-consuming), and a high capital cost.

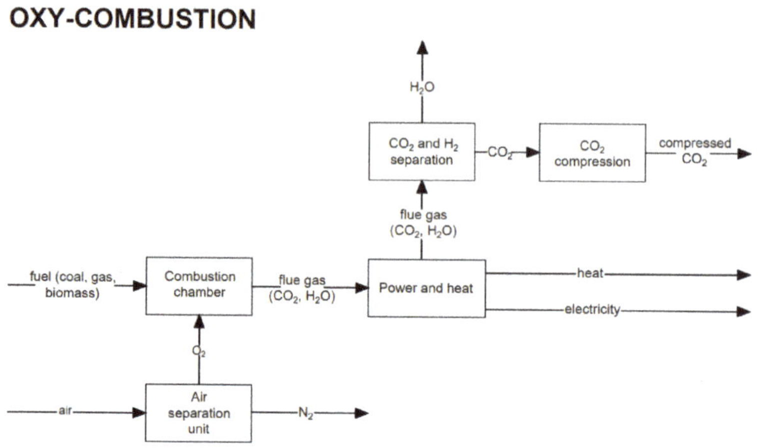

OXY-COMBUSTION

H_2O

| CO_2 and H_2 separation | —CO_2→ | CO_2 compression | compressed CO_2 → |

flue gas (CO_2, H_2O)

fuel (coal, gas, biomass) → Combustion chamber — flue gas (CO_2, H_2O) → Power and heat — heat → / electricity →

O_2

air → Air separation unit — N_2 →

Figure 9. Block diagram of electricity generation heat production with the oxy-combustion CO_2 capture method.

Oxy-combustion methods are mainly used at the laboratory scale, and for pilot installations, among which are the Callide Power Station [51] and Compostilla Thermal Power [52]. A 30 MW_e experimental unit in the Callide Power Station started operation in 2012. The nominal flow of 98% pure oxygen, which was supplied to the boiler, was 19,200 m_n^3/h. Various types of fuel have been the object of investigation (Callide coal, Minerva coal, etc.). Daily production of CO_2 based on cryogenic capture technology was 75 t. The unit was closed after a successful demonstration in 2016. A pilot power plant unit with CO_2 capture, based on oxy-combustion technology, was developed as part of the OXYCFB 300 project at Compostilla Thermal Power. The CO_2 capture installation was equipped with a circulating fluidized bed boiler with thermal power of 30 MW_{th}, and a pulverized coal-fired boiler with thermal power of 20 MW_{th}. The flue gas stream was 800 m^3/h. The daily capacity of the separation process, which was carried out using the cryogenic method, was 3–5 tons of CO_2.

The development of oxy-combustion technology is not only connected with solid-fired fuel units. It also concerns systems equipped with gas turbines, where combustion in an oxygen-enriched atmosphere takes place. Within the framework of the project Negative CO_2 Emission Gas Power Plant [53,54], the concept of a negative CO_2 emission gas power plant based on oxy-combustion combined with CO_2 capture from flue gas was developed. Application of CO_2 neutral fuel, such as sewage sludge in combination with oxy-combustion and CO_2 capture, will allow negative emission levels [55]. The CO_2 capture process will be possible, among other things, through the use of a prototype spray-ejector condenser (SEC). The main task of the SEC will be to condense the water vapor from the exhaust gases.

Among other CO_2 capture technologies, oxy-combustion carbon capture does not require many process modifications. Oxygen is used for the combustion process instead of air to eliminate the nitrogen content in flue gas, which leaves flue gas with carbon dioxide and water vapor as the major contents. This flue gas does not require high energy consumption for CO_2 separation. The boiler's temperature is controlled by again sending part of the flue gas (about 70%) to the boiler. The air separation unit (ASU), which separates oxygen from air, is the most energy-consuming part of the oxyfuel combustion process. An industrial-scale cryogenic ASU consumes up to 200–225 kWh/t on an industrial scale [56]. To improve efficiency and CO_2 capture, oxyfuel combustion carbon capture is combined with moderate and intense low-oxygen dilution (MILD). MILD oxy-combustion carbon capture (MOFC) holds many benefits, such as improving the efficiency of the plant, improving the purity of the CO_2, and reducing energy consumption [12]. The oxy-fuel combustion CCS technology used in the Allam cycle shows a higher efficiency of 55–59%, which is higher when compared to that of a combined cycle power plant with a carbon capture unit. In the Allam cycle, the heat generated from the ASU is sent to the regenerator to heat up the CO_2 to 400 °C, which is then reused in the combustion chamber, improving the cycle efficiency [57].

5. Indicators for CO_2 Emission Level Assessment

In order assess the CO_2 emission level and CO_2 capture, many variants of indicators, depending on what they want to present, can be used and have been presented [8,58–63]. The most popular indicators that allow evaluating CO_2 technologies are as follows:

- Specific emission of carbon dioxide, eCO_2 (kg_{CO2}/kWh):

$$eCO_2 = \frac{\dot{m}_{CO_2}}{N_{net}} \cdot 3600 \tag{1}$$

where $\dot{m}_{CO_2}$—mass flow rate of the emitted CO_2, kg/s; N_{net}—net power of electricity generation, kW.

- Relative emissivity of carbon dioxide, e_rCO_2 (kg_{CO2}/kWh):

$$e_rCO_2 = \eta_{net} \cdot eCO_2 \tag{2}$$

$$e_rCO_2 = \frac{N_{net}}{\dot{Q}_{CC}} \frac{\dot{m}_{CO2}}{N_{net}} \cdot 3600 \tag{3}$$

$$e_rCO_2 = \frac{\dot{m}_{CO2}}{\dot{Q}} \cdot 3600 \tag{4}$$

where e_rCO_2 in (Equations (2)–(4)) is defined as the amount of emitted CO_2 divided by heat input from the fuel (kg_{CO2}/kWh); or as the net efficiency of electricity production of the cycle, η_{net} ($\eta_{net}=N_{net}/\dot{Q}$), multiplied by the specific CO_2 emission, eCO_2 ($eCO_2 = \dot{m}_{CO2}/N_{net}$). $\dot{Q}$ is the chemical energy rate, kW.

- CO_2 capture ratio CCR (unitless):

$$CCR = \frac{\dot{m}_{CO_2,capt}}{\dot{m}_{CO_2,gen}} \tag{5}$$

where CCR (unitless) is defined as the mass flow rate of the captured CO_2, $\dot{m}_{CO_2,capt}$ (t/h) divided by the generated mass flow rate of CO_2, $\dot{m}_{CO_2,gen}$ (t/h).

- CO_2 emission index, χ (kg_{CO2}/kJ):

$$\chi = \frac{m_{CO_2,gen}}{Q} \tag{6}$$

Equation (6) defines new factor, χ (kg_{CO2}/kJ), which is the amount of CO_2 mass generated (kg) to the heat input in the fuel (kJ). $m_{CO_2,gen}$ is the mass of generated CO_2, kg; Q is the heat input by the fuel, kJ.

- CO_2 captured (kg_{CO2}/kWh):

$$CO_2 \; captured = \frac{\chi}{\eta_{p,CCS}} \eta_{cap} \tag{7}$$

$$CO_2 \; captured = \frac{m_{CO_2,gen}}{Q} \frac{\eta_{cap}}{\eta_{p,CCS}} \tag{8}$$

The term CO_2 *captured* (kg_{CO2}/kWh), defined by Equations (7) and (8), refers to the amount of CO_2 *captured* (kg_{CO2}/kJ) per unit of main product of the plant (e.g., power in power plant, kWh); $\eta_{p,CCS}$ is the efficiency of the plant with capture; η_{cap} is the efficiency of the CO_2 capture.

- CO_2 *emitted* (kg_{CO2}/kWh):

$$CO_2 \; emitted = \frac{\chi}{\eta_{p,CCS}}\left(1 - \eta_{cap}\right) \tag{9}$$

The term CO_2 *emitted* (kg_{CO2}/kWh) is specified as the amount of CO_2 *emitted* per main product of the plant (e.g., power in the power plant, kWh).

- CO_2 *avoided* (kg_{CO2}/kWh):

$$CO_2 \; avoided = \frac{eCO_{2ref} - eCO_{2p,CCS}}{eCO_{2ref}} \tag{10}$$

The indicator CO_2 *avoided* (-) evaluates the direct CO_2 emission reduction from the plant, taking into account the emissions related to the capture processes, e.g., steam generation, and the emissions of the flue gas. eCO_{2ref} is the specific emission from the reference plant (kg_{CO2}/kWh), and $eCO_{2p,CCS}$ is the specific emission from the plant with capture (kg_{CO2}/kWh).

The term CO_2 *avoided* can also be characterized by the following form:

$$CO_2 \; avoided = \frac{\chi}{\eta_{ref}} - \frac{\chi}{\eta_{p,CCS}}\left(1 - \eta_{cap}\right) \tag{11}$$

This parameter is specified as the net reduction of CO_2 emission per unit of net power output (kg_{CO2}/kWh) comparing with a reference power plant without CO_2 capture and compared to that of a similar power plant with CO_2 capture. η_{ref} is the efficiency of the reference plant without capture, $\eta_{p,CCS}$ is the efficiency of the reference plant with CO_2 capture installation, and η_{cap} is the efficiency of the CO_2 capture process.

- Specific primary energy consumption for CO_2 avoided $SPECCA$ ($kWh/kgCO_2$):

$$SPECCA = \frac{HR_{p,CCS} - HR_{ref}}{eCO_{2ref} - eCO_{2p,CCS}} = \frac{\frac{1}{\eta_{p,CCS}} - \frac{1}{\eta_{ref}}}{eCO_{2ref} - eCO_{2p,CCS}} \tag{12}$$

The indicator from Equation (12) defines the amount of energy used to avoid 1 kg of emitted CO_2 (kWh/kg_{CO2}). $HR_{p,CCS}$ and HR_{ref} (kJ/kWh) are the heat rate of the plant with and without CO_2 capture, respectively.

- Specific primary energy consumption cost for CO_2 avoided (€/$kgCO_2$):

$$SPECCA_{cost} = \frac{HR_{p,CCS} - HR_{ref}}{eCO_{2ref} - eCO_{2p,CCS}} EC = \frac{\frac{1}{\eta_{p,CCS}} - \frac{1}{\eta_{ref}}}{eCO_{2ref} - eCO_{2p,CCS}} EC \tag{13}$$

Equation (13) is defined as the product of the $SPECCA$ index multiplied by the primary energy cost, (€/$kgCO_2$), where HR_{ref} and $HR_{p,CCS}$ are the heat rates of the plant without and with CCS installation, respectively (kJ/kWh); eCO_{2ref} and $eCO_{2p,CCS}$ are the CO_2 emission rates in the plant without and with CCS installation, respectively (kg_{CO2}/kWh); EC is the primary energy cost (€/kWh).

- Levelized costs of electricity (USD/MWh):

$$LCOE = \frac{\sum (Capital_t + O\&M_t + Fuel_t + Carbon_t + D_t) \cdot (1 + r)^{-t}}{\sum MWh (1 + r)^{-t}} \tag{14}$$

The levelized costs of electricity (*LCOE*), according to IEA, indicates the economic costs of generic technology. It allows comparing technology over operating lifetimes at plant-level unit costs, at different baseloads. Equation (11) calculates the average lifetime levelized costs based on the costs of construction, operation and maintenance, fuel, carbon emissions, and decommissioning and dismantling, where $Capital_t$ is the total capital construction costs in year (t), $O\&M_t$ is the operation and maintenance costs in year t, $Fuel_t$ is the fuel costs in year t, $Carbon_t$ is thecarbon costs in year t, D_t is the decommissioning and waste management costs in year t, MWh is the amount of the electricity produced annually (MWh), $(1+r)^{-t}$ is the real discount rate corresponding to the cost of capital, and the subscript t means the year, when is a sale of production or takes place the cost disbursement.

6. Applications of CO_2 Capture Technologies on a Large-Industrial Scale

Carbon capture, utilization, and storage (CCUS) will have a key role in efforts that will lead the world to a net-zero CO_2 emission path. CCUS technologies will have to play an important role, alongside electrification, hydrogen technologies, and sustainable energy based on biofuels. It is the only group of technologies that directly contributes to a reduction in CO_2 in crucial sectors and CO_2 removal that cannot be avoided. Stronger climate and investment incentives are driving forces for CCUS technology. Since 2017, the rapid growth of newly announced integrated CCUS units has been observed (Figure 10). These are mainly located in the United States and Europe and in Australia, China, Korea, the Middle East, and New Zealand [6]. If all projects are launched, the global CO_2 capture potential will more than triple to about 130–150 Mt_{CO2}/year of captured CO_2 (currently it is about 40 Mt_{CO2}/year), as shown in Figure 10. In 2021, 97 CCUS facilities were in early stages and announced, 66 were in advanced development and 5 were under construction [64].

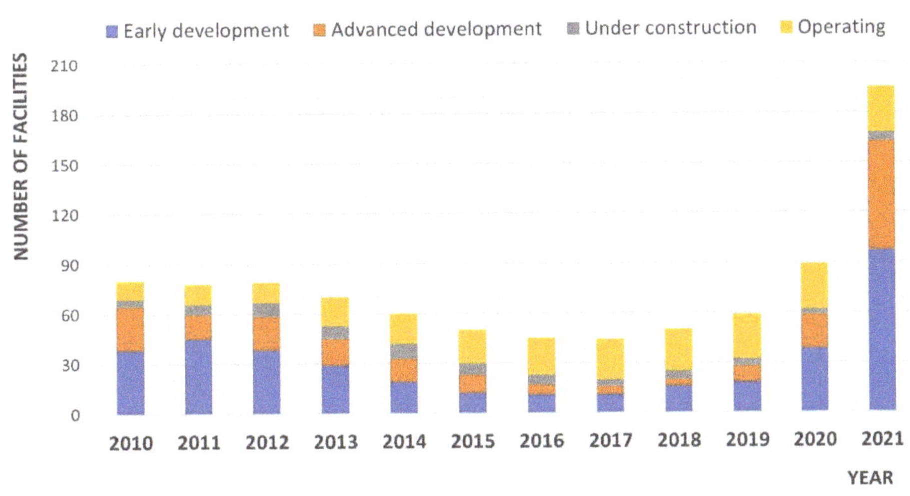

Figure 10. Commercial CCS facilities operating and under development [6,64–67].

Currently, there are 27 CCUS facilities in the world with a capture capacity of up to 40 Mt CO_2 per year [6,64]. The total capacity of all CCUS installations developed since 1972, and the capacity of new installations built every year, are presented in Figure 11.

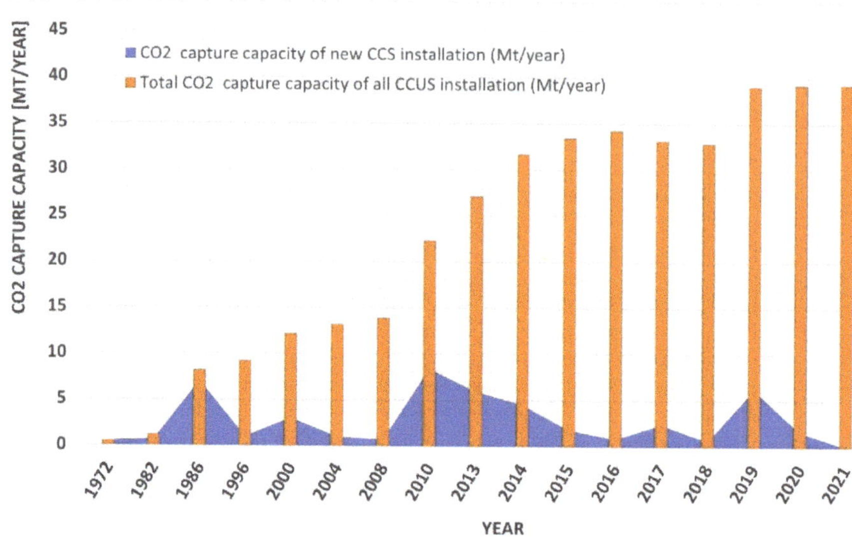

Figure 11. Total capacity and new large-scale CCUS installations capacities in 1972–2021 [6,64,66–70].

Some of these have been operating since the 1970s and 1980s when natural gas processing plants in Texas began capturing CO_2 and delivering it to local oil producers. The first large-scale CO_2 capture and injection project, with dedicated CO_2 storage and monitoring systems, was put into operation at the undersea Sleipner gas field in Norway [66,67] in 1965. Commercial large-scale operations in 2021 in terms of CO_2 capture facilities are presented in Table 2. Commercial large-scale is defined as a scale covering the capture of at least 0.8 Mt/year CO_2 for coal-fired power stations and 0.4 Mt/year CO_2 for other industrial facilities (including natural gas-fired power generation). Data presented in Figure 11 include facilities that are out of service: that in Salah (Algeria) closed in 2011, Los Cabin Gas Plant (USA) closed in 2018, Kemper County IGCC Project (Canada) closed in 2017, and Petra Nova (USA) closed in 2020.

Table 2. Commercial large-scale CCUS facilities in operation between 1972 and 2021 ([1] facility out of operation in 2021) [6,66–70].

Country	Project	Operation Date	Source of CO_2	CO_2 Capture Capacity (Mt/Year)	Primary Storage Type
United States (USA)	Terrell natural gas plants (Val Verde Gas Plants)	1972	Natural gas processing	0.4–0.5	United States (USA)
USA	Enid Fertilizer	1982	Fertilizer production	0.7	EOR
USA	Shute Creek gas processing facility	1986	Natural gas processing	7.0	EOR
Norway	Sleipner CO_2 storage project	1996	Natural gas processing	1.0	Deep saline formation
USA/Canada	Great Plains Synfuels (Weyburn/Midale)	2000	Synthetic natural gas	3.0	EOR
Algeria [1]	In Salah CO_2 Injection	2004	Natural gas processing	1.0	Deep saline formation
Norway	Snohvit CO_2 storage project	2008	Natural gas processing	0.7	Deep saline formation
USA	Century plant	2010	Natural gas processing	8.4	EOR
USA	Air Products steam methane reformer	2013	Hydrogen production	1.0	EOR
USA [1]	Lost Cabin Gas Plant	2013	Natural gas processing	0.9	EOR
USA	Coffeyville Gasification	2013	Fertilizer production	1.0	EOR
Brazil	Petrobras Santos Basin pre-salt oilfield CCS	2013	Natural gas processing	3.0	EOR
Canada	Boundary Dam CCS	2014	Power generation (coal)	1.0	EOR
Canada [1]	Kemper County IGCC Project	2014	Natural gas processing	3.5	EOR
Saudi Arabia	Uthmaniyah CO_2 EOR demonstration	2015	Natural gas processing	0.8	EOR
Canada	Quest	2015	Hydrogen production	1.0	Deep saline formation
United Arab Emirates	Abu Dhabi CCS	2016	Iron and steel production	0.8	EOR
USA [1]	Petra Nova	2017	Power generation (coal)	1.4	EOR
USA	Illinois industrial	2017	Ethanol production	1.0	Deep saline formation
China	Jilin oilfield CO_2 EOR	2018	Natural gas processing	0.6	EOR
Australia	Gorgon Carbon Dioxide Injection	2019	Natural gas processing	3.4–4.0	Deep saline formation
Qatar	Qatar LNG CCS	2019	Natural gas processing	2.2	Dedicated geological storage
Canada	Alberta Carbon Trunk Line (ACTL) with North West Redwater Partnerships	2020	Hydrogen production	1.3–1.6	EOR

7. Conclusions

Growing energy demands are still observed, and a large part of energy is produced with the use of fossil fuels. The gaseous pollutant emissions from fossil fuel power plants include carbon dioxide, which is the major cause for the emission causing global warming and climate change. The Paris Agreement aims for sustainable energy with zero-emission by capturing CO_2 released into the atmosphere from anthropogenic activities. The International Energy Agency report in 2020 recommends that the global energy transition can be carried using renewable energy, bioenergy, green hydrogen, and CCUS to reduce emissions in large-scale industries. This paper presents developed methods and technologies for carbon dioxide capture. Crucial issues connected with the progress of contemporary global technologies based on pre-combustion, post-combustion, and oxy-combustion methods have been discussed.

Pre-combustion capture is connected with removing carbon compounds before introducing fuel into the combustion chamber. This method is mainly used in integrated gasification in combined cycle (IGCC) processes, and can achieve a high efficiency of CO_2 removal—more than 90% of CO_2 capture.

Post-combustion methods are the only solutions for existing coal-fired units. CO_2 capture from flue gases is based on chemical absorption, physical separation, membrane separation, cryogenic methods, and combustion in a chemical loop. MEA solvents are the most mature technology; however, research showed less energy consumption for ammonia and PZ-AMP solvent, which must be investigated more.

In oxy-combustion technologies, fuel is burnt in an oxygen-enriched atmosphere. Therefore, exhaust gases consist mainly of CO_2 and steam, which are then condensed and carbon dioxide is separated. Currently, oxy-combustion methods are used at the laboratory scale and in pilot installations. This technology has many advantages and opportunities for development (reduction of NO_x, less amount of exhaust gases, does not require many process modifications, possibility to use CO_2 neutral fuel, and Allam cycle).

Different ways of reducing CO_2, when the fossil fuels are used as the energy input, are presented in this paper:

- In the case of fossil-fueled power plants, there is a need to use carbon capture utilization and storage methods to reduce CO_2 emissions, and, at the end, to minimize the impact of greenhouse gases on the environment.
- Currently, there are 27 CCUS commercial facilities with which the global CO_2 capture potential is about 40 $MtCO_2$/year, but this can increase by three times after launching announced CCUS units.
- Based on the prepared review, it can be concluded that most of the operating, large-scale, commercial CCUS facilities are connected with natural gas processing and use CO_2 to enhance oil recovery.

Reduction of CO_2 emissions in energy technologies, especially in high-power fossil-fueled technologies, requires constant development in order to achieve relevant capacities. The indicators for CO_2 emission level assessment, as specific emissions of CO_2, CCR, CO_2 *avoided*, or *SPECCA*, are helpful in the evaluation process of different developed technologies for CO_2 capture and were described in detail in this paper.

Author Contributions: Conceptualization, P.M.; methodology, P.M., K.C., N.S. and T.K. formal analysis, P.M., K.C.; investigation, P.M., K.C., N.S. and T.K.; resources, P.M., K.C., N.S. and T.K.; writing—original draft preparation, P.M., K.C., N.S. and T.K.; writing—review and editing, P.M., K.C., N.S. and T.K.; visualization, P.M., K.C., N.S. and T.K.; supervision, P.M.; project administration, P.M.; funding acquisition, P.M. All authors have read and agreed to the published version of the manuscript.

Funding: The research leading to these results has received funding from the Norway Grants 2014–2021 via the National Centre for Research and Development. Article has been prepared within the frame of the project: "Negative CO_2 emission gas power plant"—NOR/POLNORCCS/NEGATIVE-CO2-PP/0009/2019-00 which is co-financed by program "Applied research" under the Norwegian Financial Mechanisms 2014–2021 POLNOR CCS 2019—Development of CO_2 capture solutions integrated in power and industry processes.

Institutional Review Board Statement: Not applicable.

Informed Consent Statement: Not applicable.

Data Availability Statement: Not applicable.

Conflicts of Interest: The authors declare no conflict of interest. The funders had no role in the design of the study; in the collection, analyses, or interpretation of data; in the writing of the manuscript, or in the decision to publish the results.

References

1. Spliethoff, H. *Power Generation from Solid Fuels*; Springer: Berlin/Heidelberg, Germany, 2010.
2. Chmielniak, T. *Technologie Energetyczn*, 2nd ed.; Wydawnictwo Naukowe PWN: Warszawa, Poland, 2021.
3. Sarkar, S. *Fuels and Combustion*; University Press: New Delhi, India, 2009.
4. Madejski, P. (Ed.) *Thermal Power Plants, New Trends and Recent Developments*; Intech Open Limited: London, UK, 2018; ISBN 978-1-78923-079-6. [CrossRef]
5. Madejski, P.; Janda, T.; Modliński, N.; Nabagło, D. A Combustion Process Optimization and Numerical Analysis for the Low Emission Operation of Pulverized Coal-Fired Boiler. In *Developments in Combustion Technology*; Kyprianidis, K., Skvaril, J., Eds.; In Intech Open Limited: London, UK, 2016. [CrossRef]
6. *Energy Technology Perspectives 2020, Special Report on Carbon Capture Utilisation and Storage CCUS in Clean Energy Transitions*; IEA: Paris, France, 2020; Available online: https://iea.blob.core.windows.net/assets/7f8aed40-89af-4348-be19-c8a67df0b9ea/Energy_Technology_Perspectives_2020_PDF.pdf (accessed on 2 December 2021).
7. Centrum Informacji o Rynku Energii, Cena Emisji CO2 Może Wzrosnąć o Ponad 50% do 2030—Wynika z Projektu UE. Available online: https://www.cire.pl/artykuly/materialy-problemowe/186670-cena-emisji-co2-moze-wzrosnac-o-ponad-50-do-2030-wynika-z-projektu-ue (accessed on 20 October 2021).
8. Nord, L.; Bolland, O. *Carbon Dioxide Emission Management in Power Generation*; Wiley-VCH Verlag GmbH & Co.: Weinheim, Germany, 2020.
9. Peridas, G.; Mordick Schmidt, B. The role of carbon capture and storage in the race to carbon neutrality. *Electr. J.* **2021**, *34*, 106996. [CrossRef]
10. Wienchol, P.; Szlęk, A.; Ditaranto, M. Waste-to-energy technology integrated with carbon capture—Challenges and opportunities. *Energy* **2020**, *198*, 117352. [CrossRef]
11. Holz, F.; Scherwath, T.; del Granado, P.C.; Skar, C.; Olmos, L.; Ploussard, Q.; Ramos, A.; Herbst, A. A 2050 perspective on the role for carbon capture and storage in the European power system and industry sector. *Energy Econ.* **2021**, *104*, 105631. [CrossRef]
12. Gładysz, P.; Stanek, W.; Czarnowska, L.; Sładek, S.; Szlęk, A. Thermo-ecological evaluation of an integrated MILD oxy-fuel combustion power plant with CO_2 capture, utilisation, and storage—A case study in Poland. *Energy* **2018**, *144*, 379–392. [CrossRef]
13. Shijaz, H.; Attada, Y.; Patnaikuni, V.S.; Vooradi, R.; Anne, S.B. Analysis of integrated gasification combined cycle power plant incorporating chemical looping combustion for environment-friendly utilization of Indian coal. *Energy Convers. Manag.* **2017**, *151*, 414–425. [CrossRef]
14. Mukherjee, S.; Kumar, P.; Yang, A.; Fennell, P. Energy and exergy analysis of chemical looping combustion technology and comparison with pre-combustion and oxy-fuel combustion technologies for CO_2 capture. *J. Environ. Chem. Eng.* **2015**, *3*, 2104–2114. [CrossRef]
15. Majchrzak-Kucęba, I.; Wawrzyńczak, D.; Zdeb, J.; Smółka, W.; Zajchowski, A. Treatment of flue gas in a CO_2 capture pilot plant for a commercial CFB boiler. *Energies* **2021**, *14*, 2458. [CrossRef]
16. Popa, A.; Edwards, R.; Aandi, I. Carbon capture considerations for combined cycle gas turbine. *Energy Procedia* **2011**, *4*, 2315–2323. [CrossRef]
17. Tramošljika, B.; Blecich, P.; Bonefačić, I.; Glažar, V. Advanced ultra-supercritical coal-fired power plant with post-combustion carbon capture: Analysis of electricity penalty and CO_2 emission reduction. *Sustainability* **2021**, *13*, 801. [CrossRef]
18. Theo, W.L.; Lim, J.S.; Hashim, H.; Mustaffa, A.A.; Ho, W.S. Review of pre-combustion capture and ionic liquid in carbon capture and storage. *Appl. Energy* **2016**, *183*, 1633–1663. [CrossRef]
19. Olabi, A.G.; Obaideen, K.; Elsaid, K.; Wilberforce, T.; Sayed, E.T.; Maghrabie, H.M.; Abdelkareem, M.A. Assessment of the pre-combustion carbon capture contribution into sustainable development goals SDGs using novel indicators. *Renew. Sustain. Energy Rev.* **2022**, *153*, 1117102022. [CrossRef]
20. Wu, F.; Dellenback, P.A.; Fan, M. Highly efficient and stable calcium looping based pre-combustion CO_2 capture for high-purity H2 production. *Mater. Today Energy* **2019**, *13*, 233–238. [CrossRef]

21. Grande, C.A.; Blom, R.; Andreassen, K.A.; Stensrød, R.E. Experimental Results of Pressure Swing Adsorption (PSA) for Pre-combustion CO_2 Capture with Metal Organic Frameworks. *Energy Procedia* **2017**, *114*, 2265–2270. [CrossRef]

22. Wanga, Y.; Zhaoa, L.; Ottoa, A.; Robiniusa, M.; Stoltena, D. A Review of Post-combustion CO_2 Capture Technologies from Coal-fired Power Plants. *Energy Procedia* **2017**, *114*, 650–665. [CrossRef]

23. Kuropka, J. Możliwości Ograniczania Emisji Ditlenku Węgla ze Spalin Energetycznych. Available online: http://www.pzits.not.pl/docs/ksiazki/Pol_%202012/Kuropka%20179-188.pdf (accessed on 18 October 2021).

24. Vega, F.; Cano, M.; Camino, S.; Fernandez, L.M.G.; Portillo, E.; Navarrete, B. Solvents for Carbon Dioxide Capture. In *Carbon Dioxide Chemistry, Capture and Oil Recovery*; Intechopen: London, UK, 2018; ISBN 978-1-78923-575-3. [CrossRef]

25. Artanto, Y.; Jansen, J.; Pearson, P.; Puxty, G.; Cottrell, A.; Meuleman, E.; Feron, P. Pilot-scale evaluation of AMP/PZ to capture CO2 from flue gas of an Australian brown coal–fired power station. *Int. J. Greenh. Gas Control* **2014**, *20*, 189–195. [CrossRef]

26. Boot-Handford, M.; Abanades, J.C.; Anthony, E.J.; Blunt, M.J.; Brandani, S.; Mac Dowell, N.; Fernández, J.R.; Ferrari, M.-C.; Gross, R.; Hallett, J.P.; et al. Carbon capture and storage update. *Energy Environ. Sci.* **2014**, *7*, 130–189. [CrossRef]

27. Bui, M.; Adjiman, C.S.; Bardow, A.; Anthony, E.J.; Boston, A.; Brown, S.; Fennell, P.S.; Fuss, S.; Galindo, A.; Hackett, L.A.; et al. Carbon capture and storage (CCS): The way forward. *Energy Environ. Sci.* **2018**, *11*, 1062–1176. [CrossRef]

28. Barzgali, F.; Lai, S.; Mani, F. Novel non-aqueous amine solvents for reversible CO_2 capture. *Energy Procedia* **2014**, *63*, 1795–1804. [CrossRef]

29. Mahi, M.; Mokbel, I.; Negadi, L.; Dergal, F.; Jose, J. Experimental solubility of carbon dioxide in monoethanolamine, or diethanolamine or N-methyldiethanolamine (30 wt%) dissolved in deep eutectic solvent (choline chloride and ethylene glycol solution). *J. Mol. Liq.* **2019**, *289*, 111062. [CrossRef]

30. Sifat, N.S.; Haseli, Y. A critical review of CO2 Capture Technologies and Prospects for Clean Power Generetion. *Energies* **2019**, *12*, 4143. [CrossRef]

31. Osman, A.I.; Hefny, M.; Maksoud, M.I.A.A.; Elgarahy, A.M.; Rooney, D.W. Recent advances in carbon capture storage and utilisation technologies: A review. *Environ. Chem. Lett.* **2020**, *19*, 797–849. [CrossRef]

32. Li, J.-R.; Ma, Y.; McCarthy, M.C.; Sculley, J.; Yu, J.; Jeong, H.-K.; Balbuena, P.B.; Zhou, H.-C. Carbon dioxide capture-related gas adsorption and separation in metal-organic frameworks. *Coord. Chem. Rev.* **2011**, *255*, 1791–1823. [CrossRef]

33. Pires, J.C.M.; Martins, F.G.; Alvim-Ferraz, M.C.M.; Simões, M. Recent developments on carbon capture and storage: An overview. *Chem. Eng. Res. Des.* **2011**, *89*, 1446–1460. [CrossRef]

34. Kárászová, M.; Zach, B.; Petrusová, Z.; Červenka, V.; Bobák, M.; Šyc, M.; Izák, P. Post-combustion carbon capture by membrane separation, Review. *Sep. Purif. Technol.* **2020**, *238*, 116448. [CrossRef]

35. Rakowski, J.; Bocian, P.; Celińska, A.; Świątkowski, B.; Golec, T. Zastosowanie Pętli Chemicznej w Energetyce. Available online: http://elektroenergetyka.pl/upload/file/2016/4/Rakowski_04_2016.pdf (accessed on 15 October 2021).

36. Bhavsar, S.; Najera, M.; More, A.; Veser, G. Chemical-looping processes for fuel-flexible combustion and fuel production. In *Reactor and Process Design in Sustainable Energy Technology*, 1st ed.; Elsevier B.V.: Amsterdam, The Netherlands, 2014; pp. 233–280. [CrossRef]

37. Tilak, P.; El-Halwagi, M.M. Process integration of Calcium Looping with industrials plants for monetizing CO_2 into value-added products. *Carbon Resour. Convers.* **2018**, *1*, 191–199. [CrossRef]

38. Knapik, E.; Kosowski, P.; Stopa, J. Cryogenic liquefaction and separation of CO_2 using nitrogen removal unit cold energy. *Chem. Eng. Res. Des.* **2018**, *131*, 66–79. [CrossRef]

39. Song, C.; Liu, Q.; Deng, S.; Li, H.; Kitamura, Y. Cryogenic-based CO_2 capture technologies: State-of-the-art developments and current challenges. *Renew. Sustain. Energy Rev.* **2019**, *101*, 265–278. [CrossRef]

40. Mostafavi, E.; Ashrafi, O.; Navarri, P. Assessment of process modifications for amine-based post-combustion carbon capture processes. *Clean. Eng. Technol.* **2021**, *4*, 100249. [CrossRef]

41. Chao, C.; Deng, Y.; Dewil, R.; Baeyens, J.; Fan, X. Post-combustion carbon capture. *Renew. Sustain. Energy Rev.* **2021**, *138*, 110490. [CrossRef]

42. Lungkadee, T.; Onsree, T.; Tangparitkul, S.; Janwiruch, N.; Nuntaphan, A.; Tippayawong, N. Technical and economic analysis of retrofitting a post-combustion carbon capture system in a Thai coal-fired power plant. *Energy Rep.* **2021**, *7*, 308–313. [CrossRef]

43. Otitoju, O.; Oko, E.; Wang, M. Technical and economic performance assessment of post-combustion carbon capture using piperazine for large scale natural gas combined cycle power plants through process simulation. *Appl. Energy* **2021**, *292*, 116893. [CrossRef]

44. El Hadri, N.; Quang, D.V.; Goetheer, E.L.V.; Abu Zahra, M.R.M. Aqueous amine solution characterization for post-combustion CO_2 capture process. *Appl. Energy* **2017**, *185*, 1433–1449. [CrossRef]

45. Zhang, Z.; Pan, S.-Y.; Li, H.; Cai, J.; Olabi, A.G.; Anthony, E.J.; Manovic, V. Recent advances in carbon dioxide utilization. *Renew. Sustain. Energy Rev.* **2020**, *125*, 109799. [CrossRef]

46. Agarwal, A.S.; Rode, E.; Sridhar, N.; Hill, D. Conversion of CO2 to Value-Added Chemicals: Opportunities and Challenges. In *Handbook Climate Change Mitigation and Adaptation*; Chen, W.Y., Suzuki, T., Lackner, M., Eds.; Springer: New York, NY, USA, 2015; pp. 1–40. [CrossRef]

47. Pal, T.K.; De, D.; Bharadwaj, P.K. Metal–organic frameworks for the chemical fixation of CO_2 into cyclic carbonates. *Coord. Chem. Rev.* **2020**, *408*, 213173. [CrossRef]

48. Kumaravel, V.; Barlett, J.; Pillai, S.C. Photoelectrochemical Conversion of Carbon Dioxide (CO_2) into Fuels and Value Added Products. *ACS Energy Lett.* **2020**, *5*, 486–529. [CrossRef]

49. McGrath, O.J. Biological Conversion of Carbon Dioxide to Value-Added Chemicals. *Grad. Thesis Diss. Probl. Rep.* **2021**, 8293. [CrossRef]

50. Nocito, F.; Dibenedetto, A. Atmospheric CO_2 mitigation technologies: Carbon capture utilization and storage. *Curr. Opin. Green Sustain. Chem.* **2020**, *21*, 34–43. [CrossRef]

51. CS Energy, Callide Oxyfuel Project. Available online: https://www.csenergy.com.au/what-we-do/generating-energy/callide-power-station/callide-oxyfuel-project (accessed on 20 October 2021).

52. Global CCS Institue. The Compostilla Project OXYCFB300. Available online: https://www.globalccsinstitute.com/archive/hub/publications/137158/Compostilla-project-OXYCFB300-carbon-capture-storage-demonstration-project-knowledge-sharing-FEED-report.pdf (accessed on 20 October 2021).

53. Project Negative CO2 Emission Gas Power Plant. Available online: https://nco2pp.mech.pg.gda.pl (accessed on 20 October 2021).

54. Ziółkowski, P.; Madejski, P.; Amiri, M.; Kuś, T.; Stasiak, K.; Subramanian, N.; Pawlak-Kruczek, H.; Badur, J.; Niedźwiecki, Ł.; Mikielewicz, D. Thermodynamic Analysis of Negative CO_2 Emission Power Plant Using Aspen Plus, Aspen Hysys, and Ebsilon Software. *Energies* **2021**, *14*, 6304. [CrossRef]

55. Ziółkowski, P.; Badur, J.; Pawlak-Kruczek, H.; Niedzwiecki, L.; Kowal, M.; Krochmalny, K. A novel concept of negative CO_2 emission power plant for utilization of sewage sludge. In Proceedings of the 6th International Conference on Contemporary Problems of Thermal Engineering CPOTE 2020, Gliwice, Poland, 21–24 September 2020; pp. 531–542.

56. Cormos, C.C. Oxy-combustion of coal, lignite and biomass: A techno-economic analysis for a large scale Carbon Capture and Storage (CCS) project in Romania. *Fuel* **2016**, *169*, 50–57. [CrossRef]

57. Kindra, V.; Rogalev, A.; Lisin, E.; Osipov, S.; Zlyvko, O. Techno-economic analysis of the oxy-fuel combustion power cycles with near-zero emissions. *Energies* **2021**, *14*, 5358. [CrossRef]

58. Bonalumi, D.; Valenti, G.; Lillia, S.; Fosbol, P.L.; Thomsen, K. A layout for the Carbon Capture with Aqueous Ammonia without Salt Precipitation. *Energy Procedia* **2016**, *86*, 134–143. [CrossRef]

59. Voldsund, M.; Gardarsdottir, S.O.; De Lena, E.; Pérez-Calvo, J.-F.; Jamali, A.; Berstad, D.; Fu, C.; Romano, M.; Roussanaly, S.; Anantharaman, R.; et al. Comparison of Technologies for CO_2 Capture from Cement Production—Part 1: Technical Evaluation. *Energies* **2019**, *12*, 559. [CrossRef]

60. Roussanaly, S.; Anantharaman, R. Cost-optimal CO_2 capture ratio for membrane-based capture from different CO_2 sources. *Chem. Eng. J.* **2017**, *327*, 618–628. [CrossRef]

61. Manzolini, G.; Giuffrida, A.; Cobden, P.D.; van Dijk, H.A.J.; Ruggeri, F.; Consonni, F. Techno-economic assessment of SEWGS technology when applied to integrated steel-plant for CO_2 emission mitigation. *Int. J. Greenh. Gas Control* **2020**, *94*, 102935. [CrossRef]

62. Chiesa, P.; Campanari, S.; Manzolini, G. CO_2 cryogenic separation from combined cycles integrated with molten carbonate fuel cells. *Int. J. Hydrogen Energy* **2011**, *36*, 10355–10365. [CrossRef]

63. International Energy Agency. *Projected Costs of Generating Electricity*; OECD Publishing: Paris Cedex, France, 2020.

64. Carbon Capture, Utilization and Storage, IEA 2021. Available online: https://www.iea.org/fuels-and-technologies/carbon-capture-utilisation-and-storage (accessed on 2 December 2021).

65. An Analysis Based on GCCSI (2020), Facilities Database. Available online: https://co2re.co/FacilityData (accessed on 20 October 2021).

66. Global CCS Institute. The Global Status of CCS 2019: Targeting Climate Change. Available online: www.globalccsinstitute.com (accessed on 2 December 2021).

67. Carbon Capture & Sequestration Technologies. Available online: https://sequestration.mit.edu/tools/projects/sleipner.html (accessed on 2 December 2021).

68. Global Institute. Global Status of CCS 2021 CCS Accelerating to Net Zero. Report. Available online: https://www.globalccsinstitute.com/resources/global-status-report/ (accessed on 2 December 2021).

69. CCUS in Power. Report. IEA 2021. Available online: https://www.iea.org/reports/ccus-in-power (accessed on 2 December 2021).

70. CCUS in Industry and Transformation. IEA 2021. Available online: https://www.iea.org/reports/ccus-in-industry-and-transformation (accessed on 2 December 2021).

Article

Forecasting of Heat Production in Combined Heat and Power Plants Using Generalized Additive Models

Maciej Bujalski [1,2] and Paweł Madejski [2,*]

[1] PGE Energia Ciepła S.A., ul. Złota 59, 00-120 Warszawa, Poland; bujalski@agh.edu.pl
[2] Department of Power Systems and Environmental Protection Facilities, Faculty of Mechanical Engineering and Robotics, AGH University of Science and Technology, Mickiewicz 30 Av., 30-059 Krakow, Poland
* Correspondence: madejski@agh.edu.pl

Abstract: The paper presents a developed methodology of short-term forecasting for heat production in combined heat and power (CHP) plants using a big data-driven model. An accurate prediction of an hourly heat load in the day-ahead horizon allows a better planning and optimization of energy and heat production by cogeneration units. The method of training and testing the predictive model with the use of generalized additive model (GAM) was developed and presented. The weather data as an input variables of the model were discussed to show the impact of weather conditions on the quality of predicted heat load. The new approach focuses on an application of the moving window with the learning data set from the last several days in order to adaptively train the model. The influence of the training window size on the accuracy of forecasts was evaluated. Different versions of the model, depending on the set of input variables and GAM parameters were compared. The results presented in the paper were obtained using a data coming from the real district heating system of a European city. The accuracy of the methods during the different periods of heating season was performed by comparing heat demand forecasts with actual values, coming from a measuring system located in the case study CHP plant. As a result, a model with an averaged percentage error for the analyzed period (November–March) of less than 7% was obtained.

Keywords: heat demand prediction; generalized additive model; combined heat and power plant; district heating network; heat production

Citation: Bujalski, M.; Madejski, P. Forecasting of Heat Production in Combined Heat and Power Plants Using Generalized Additive Models. *Energies* **2021**, *14*, 2331. https://doi.org/10.3390/en14082331

Academic Editor: Dimitrios Katsaprakakis

Received: 10 March 2021
Accepted: 18 April 2021
Published: 20 April 2021

1. Introduction

District heating systems (DHS) are common forms of heat distribution in large urban areas. The largest systems consist of a district heating network (DHN) supplied by a combined heat and power (CHP) plant. In recent years, combined-cycle gas turbine plants, as well as gas turbines with heat recovery units, have become popular due to their high flexibility, short start-up time, and lower environmental impact compared to coal plants [1]. In CHP plants, the basic product is useful heat, whose demand in a DHS must be covered at any time. This means that actual electricity production depends on the current heat demand. On the other hand, dynamic changes in electricity prices are observed. On the European energy markets, the price changes every hour, with maximum daily variations of 50% and more. An accurate prediction of an hourly heat load in the day-ahead horizon allows better planning and optimization of heat and electricity production by cogeneration units.

New solutions to optimize electricity generation and keep the produced heat load on the required level taking into account economic aspects are under constant development. Fang et al. [2] proposed an optimization model, where the heat demand and electricity price forecasts are used as an input to obtain a heat storage operation plan. Wand et al. [3] studied the flexibility of two different CHP units considering the day-ahead market and real-time wind power balancing. Nowadays, the progressive development of digitization and the use of advanced data analysis methods is a trend in the so-called 4th generation district heating [4,5]. The main element is short-term (up to several days) planning of energy production, based on the expected heat load profile [6]. An important issue is that estimated production on an hourly basis must be contracted in a day-ahead electricity market. Similarly fuel such as natural gas can also be purchased on the spot market. Moreover, inaccurate estimates of expected volumes may result in the need to switch on the peak units (e.g., heat only boilers), whose energy and environmental efficiency is lower.

Prediction of an hourly heat demand on a large urban scale is a complex problem. The heat for heating of buildings depends mainly on the weather data, while domestic hot water consumption is strongly related to consumers' behavior over the day and week. An important aspect are transient DHN effects such as heat losses, thermal inertia of buildings, etc. [7]. Forecasting methods are based on a data-driven approach. It means that heat demand and its relation to predictor variables is found in historical data. Spoladore et al. [8] analyzed data of heat demand for town-level aggregation and developed a model of hourly gas consumption for heating purposes. Nigitz et al. [9] proposed a model, where changes in consumer behavior are covered by continuous adaptation by using historical data for the ambient temperature and the heat load. Mosavi et al. [10] and Bourdeau et al. [11] gave an overview of data-driven methods that can be applied to heat load forecasting. There are models on a scale of individual buildings as well as the entire district heating network. Generally, predictive models are based on the supervised learning technique and supplied with weather data such as temperature, wind speed, cloud cover, among others. Dotz-dauer [12] developed a simple model based on stepwise regression of ambient temperature. Baltputnis et al. [13] used a multiple linear regression of meteorological parameters. Autoregressive integrated moving average ARIMA method of heat load and ambient temperature time series were used by Grosswindhager et al. [14] and Fang et al. [15]. Artificial Neural Networks ANN have been shown by Wojdyga [16] to be effective approach to analyze data from previous heating seasons. In recent years, there has been a clear increase in interest in the use of machine learning methods, such as support vector machines (SVM) [17], random forest [18], deep learning [19] and gradient boosting [20]. This results from the intensive development of IT tools used in large industrial installations, database capacity, and the availability of data acquisition systems. An important issue in the data-driven models is the selection of input variables and the way of training the model. Machine learning techniques enable analyzing many variables besides weather data, such as operating data from DHN or calendar data [21,22].

This paper focuses on the use of the generalized additive model (GAM) method to develop the heat demand model in a medium-sized heating system supplied from a CHP plant. In the GAM method, the forecast variable is estimated by smoothing the input variables with functional relationships [23]. It is useful extension of the generalized linear model (GLM), able to effectively map the seasonality and non-linearity which is normally presented in the heat load data. In the literature, there are many applications of the GAM method to forecast electricity load [24]. Kim et al. [25] decomposed the load into the components on different temporal scales, related to the annual, weekly and daily cycles. The non-linear impact of ambient air temperature on the load level was incorporated with the cubic spline by Sigauke [26]. Pathak et al. [27] used GAM to forecast gas usage for buildings taking into account weather data supplemented by features such as hour of the day, day of the week, month, etc. Khamma et al. [28] interpreted the relationships between predictors using GAM and evaluate their impacts on energy consumption in office buildings.

In this paper, the GAM method was applied to build an hourly heat demand model based on the weather data as ambient temperature, solar irradiation and wind speed. The presented method was extended by the use of variable representing the hour of the day. Application of day hour variable can improve accuracy of predicted heat load, taking into account DHN behavior changes during normal and weekend days. Additionally, the optimal size of the sliding window with data selected to the calibration layer was analyzed as a function days number. Particular attention was paid to parametrization and calibration of the model in order to obtain high accuracy in the day-ahead time horizon.

2. Case Study District Heating System

The case study DHS consists of a heating network with a total length of about 260 km. The network is supplied from a CHP plant with two identical gas turbines (total 106 MWe) equipped with heat recovery units (total 238 MWt). The flexibility of heat production in the summer season is provided by a 600 MWh heat accumulator. Four heat-only-boilers are used as the peak source. The share of individual units in total heat capacity of the plant is presented in Figure 1. The presented state refers to the period when the maximum value of produced heat in the CHP is delivered to the DHN. During the whole year, the diverse scenarios of heat production are realized and values presented in Figure 1 can rich different values. The operation strategy is strongly dependent on current heat demand as well as the overhaul time of the selected unit.

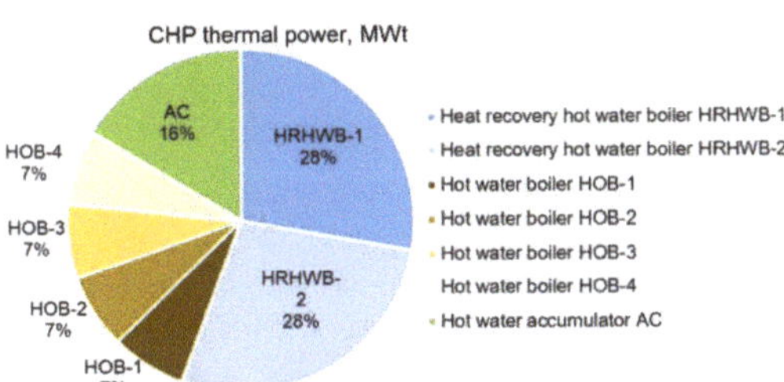

Figure 1. The structure of thermal power of the analyzed combined heat and power (CHP).

The dependence between the instantaneous heat load and the ambient air temperature during the entire heating season is depicted in Figure 2. The maximum heat load of the analyzed DHN does not exceed 250 MWt, when the minimal value of heat delivered to the final consumers during summertime is always above 20 MWt.

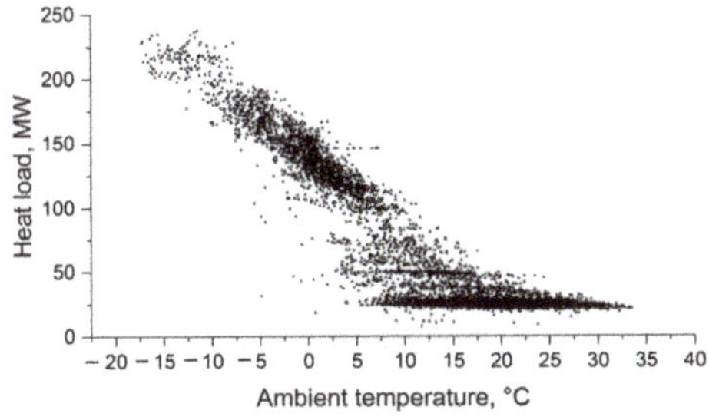

Figure 2. Heat demand variation with ambient air temperature.

Obviously, the heat load increases as the temperature drops. A strong correlation can be seen, however there is a relatively large spread for the same temperature level. Produced heat can vary in the range of around 50 MW. It results from the influence of other factors such as a month, day of the week, hour of the day and other meteorological parameters. Therefore, it is required to include additional input data in order to increase accuracy of the heat forecasts.

2.1. Physical Model of the District Heating System

The heat is supplied to the district heating network, where it is distributed to the end-users through heat distribution centers. Additional installations in the system are heating substations that connect individual sections of the network and ensure heat distribution to specific areas of the city. The instantaneous heat output transferred to the heating network by the CHP plant depends on the mass flow rate of network water and the temperature difference between supply and return side, according to Equation (1):

$$\dot{Q}_s(t) = \dot{m}(t)\, c_p\, (T_s(t) - T_R(t)) \tag{1}$$

where $\dot{Q}_s$—thermal power, kW; c_p—water specific heat, kJ/(kg·K); $\dot{m}$—mass flow rate, kg/s; T_s—supply temperature, K; T_R—return temperature, K and t—time, s

The regulation of thermal power consists of changing both the supply temperature and the mass flow rate at the output [29]. The water temperature at the outlet from the CHP plant is in accordance with the regulatory table according to Equation (2). Supply temperature depends on the ambient temperature and a coefficient depending on the wind speed and solar radiation. For example, if the weather is windy and cloudy, the coefficient c takes values above 1, increasing the supply temperature. The maximum supply temperature in the analyzed system is 115 °C, at −20 °C ambient air temperature. During summertime, the supply side is maintained at 70 °C:

$$T_s = c(W_s, I_r)\cdot f(T_a) \tag{2}$$

where: T_s—supply water temperature, K; c—coefficient; W_s—wind speed, m/s; I_r—solar radiation, W/m^2; T_a—ambient air temperature, K and f—individual system function based on the DHN regulatory table, K.

2.2. Operation of the District Heating System

During operation of the system, heat production in the source is continuously adapted to the real demand in the DHN. Due to the thermal behavior of buildings as well as transient operation of DHN there is a time delay between the current parameters at the supply and return side. The time course of parameters for selected periods during the heating season and the summer period is presented in Figures 3 and 4, respectively. There is a strong correlation of the ambient air temperature with the generated thermal power, as well as the mass flow rate of water can be observed. In the summertime, dynamic fluctuations in the water mass flow rate and heat load are visible. Outside the heating season, heat demand results from the domestic hot water consumption. Peaks associated with increased hot water consumption in the evening and morning can be observed.

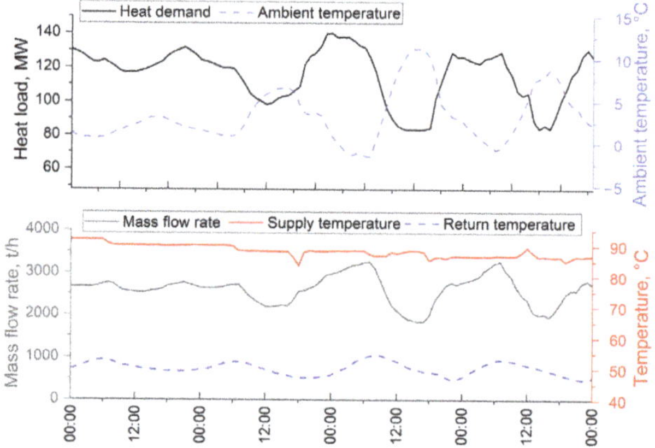

Figure 3. District Heating Network (DHN) parameters for the heating period.

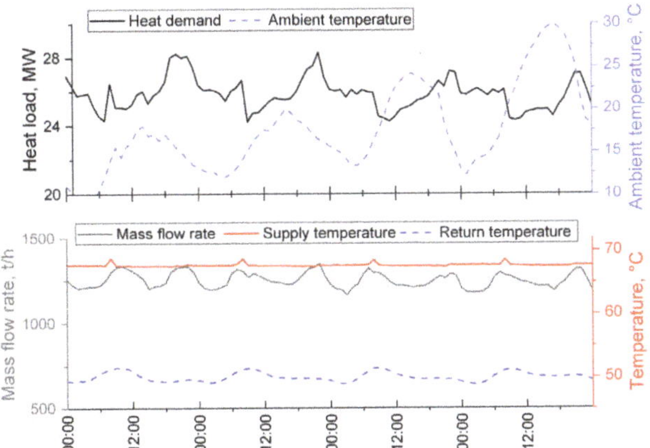

Figure 4. District heating network (DHN) parameters for summer period.

3. Heat Demand Model with the Use of GAM

Generalized Additive Model (GAM) is a class of statistical models in which the usual linear relationship between the response and predictors are replaced by several non-linear smooth functions [30]. The equation becomes as follow (Equation (3)):

$$y_i = \alpha_0 + f_1(x_{1,i}) + f_2(x_{2,i}) + \ldots + f_p(x_{p,i}) + \varepsilon_i \qquad (3)$$

where: y_i—dependent variable; α_0—intercept; $x_1, \ldots, x_p$—independent variables; $f_1, \ldots, f_p$—smoothing functions and ε_i—random error

The GAM model is able to capture the non-linear effect of individual variables. The response variable is obtained as a summation of individual effects, represented with one or more terms. The smoothing function f consists of the base functions b and the corresponding regression coefficients β (see Equation (4)). The base function b can take the form of a linear or cubic spline, P-spline, and other [31]. The smoothing function can also include two input variables, according to Equation (5), where δ is a vector of regression coefficients:

$$f(x) = \sum_{i=1}^{I} \beta_i \, b_i(x) \qquad (4)$$

$$f(x_1, x_2) = \sum_{i=1}^{I} \sum_{j=1}^{J} \delta_{ij} \, b_{1i}(x_1) \, b_{2j}(x_2) \qquad (5)$$

where: I, J—the dimension of the spline basis; $b(x)$—the corresponding spline function; β—the corresponding regression coefficient and δ—the corresponding vector of regression coefficient.

One of the advantages of GAM models is their flexibility. The method summarizes the contribution of each predictor using smoothing terms. In addition, a GAM algorithm captures nonlinearity and interactions in a learning dataset. Predictive methods with more complex mathematical approach, such as artificial neural networks (ANN) are typical black-box models. In an ANN model, interactions with a forecasted variable are created implicitly when propagating through the hidden layers as each hidden unit is a non-linear combination of the input. Besides, in order to build an accurate model with black box models, many variables, especially over a long period of time, must be taken into account [16]. In the study presented in this paper, the investigated gas-fired power plant is relatively new and there are no data from the heating seasons of previous years. For this reason, the training window is relatively short (the last several days), due to the needs of permanent adaptation to the current operation of the system, and this can be achieved by the use of GAM methods.

The heat demand model for the case study DHS was built using the *mcgv* package [32] containing the implementation of libraries with the GAM method in *R* programming language. The forecasts generated by the model were compared with real values using the root mean square error RMSE and the mean absolute percentage error MAPE (Equations (6) and (7), respectively):

$$\text{RMSE} = \sqrt{\frac{1}{N} \sum_t \left(Q_{pred,\,t} - Q_{real,\,t} \right)^2} \tag{6}$$

$$\text{MAPE} = \frac{1}{N} \sum_t \left| \frac{Q_{pred,\,t} - Q_{real,\,t}}{Q_{real,\,t}} \right| \tag{7}$$

where: N—number of hours in the analyzed period; t—hour; Q_{pred}—predicted heat load, kW and Q_{real}—real heat load, kW

$$\text{RSS} = \sum_{i=1}^{N} \left(y_i - \alpha_0 - \sum_{j=1}^{I} \beta_j \, b_j \left(x_{ij} \right) \right)^2 + \sum_{j=1}^{I} \lambda_j \int f_j''(t_j)^2 dt_j \tag{8}$$

where: λ—penalty parameter.

The model is fitted by minimalizing a penalized residual sum of squares RSS presented in Equation (8) for one dimensional basis functions. The fitting involves finding all coefficients to minimize residual sum of square with the use of the general cross validation (GCV) criteria proposed by Craven et al. [33]. The degree of smoothness in a spline can be controlled by a penalty parameter λ in order to avoid overfitting. The iterative process of minimalization is stopped when the change of value of GCV between successive iterations are less than 0.01. The function minimizing RSS provides a compromise between a regression spline fit and a linear fit. When λ is near to zero the fit will be close to the data.

3.1. Input Variables

The weather data listed in Table 1 were taken into consideration as predictors. The table contains the calculated Pearson's correlation coefficient, which is a simple statistic that measures the linear correlation between two variables. Coefficients were calculated using the data of the entire heating season. It has a value between +1 and −1, where 1 is a total positive linear correlation, and −1 is a total negative linear correlation. Correlations for forecasted and actual weather data are presented. There is less correlation with the forecasted values due to additional forecast errors. It should be noted that in a real application, the predictive model determines output based on the weather forecasts in a forthcoming day horizon.

Table 1. Weather variables with the correlation coefficient with heat demand.

Variable	Symbol	Unit	Pearson's Coefficient –Real Data	Pearson's Coefficient –Forecasted Data
Ambient temperature	*Temp*	°C	−0.81	−0.76
Solar irradiation	*Rad*	W/m^2	−0.42	−0.41
Wind speed	*Wind*	m/s	−0.20	−0.15

3.2. Flow Diagram of the Model

In Figure 5 a flow chart that includes input and output of the predictive model is presented. The model generates an hourly heat load forecast in a day-ahead time horizon, starting from 00:00. This time horizon results from the conditions related to participation in the electricity market. The model operation on the timeline diagram is illustrated in Figure 6.

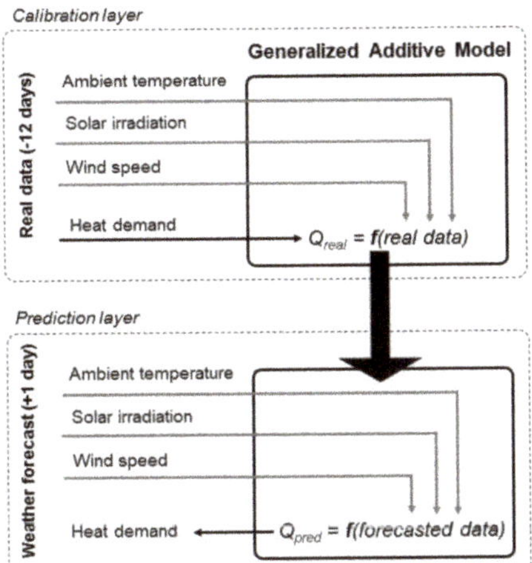

Figure 5. Flow chart of the model with calibration and prediction layer.

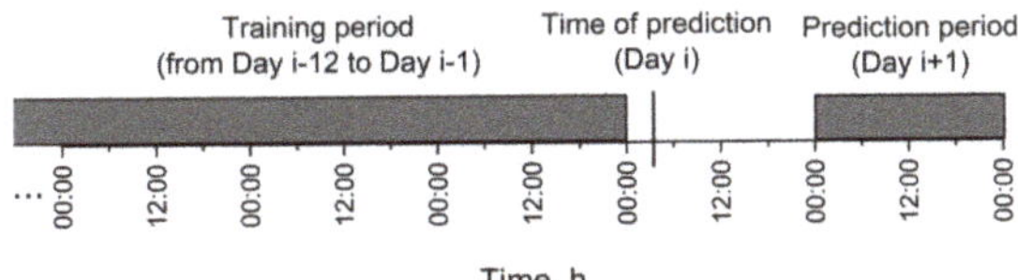

Figure 6. Timeline diagram of the heat demand model.

The model works as follows:

- Before issuing the results, the model is calibrated with the real weather data and corresponding heat output from CHP within the moving time window.
- When the weather forecast file comes at 7:00, the previously calibrated model generates load forecasts for the assumed time horizon.

3.3. Model Parametrization and Validation

Three variants of the model depending on the set of input data were considered (Equations (9)–(11)). For the first case, the model is supplied with ambient temperature, radiation and wind speed (Equation (9)). In addition to climatic parameters, variable representing the hour of the day was used, which takes values from 0 to 23. In the model M2, ambient temperate and hour of the day were implemented (Equation (10)). The model M3 only considers ambient air temperature (Equation (11)):

$$Q_{M1} = \alpha_{M1} + f_{M1}(Temp) + f_{M1}(Rad) + f_{M1}(Hour) + f_{M1}(Wind,\ Temp) \tag{9}$$

$$Q_{M2} = \alpha_{M2} + f_{M2}(Temp) + f_{M2}(Hour) \tag{10}$$

$$Q_{M3} = \alpha_{M3} + f_{M3}(Temp) \tag{11}$$

where: Q_{M1}, Q_{M2}, Q_{M3}—forecasted heat load from model M1, M2 and M3, respectively, MW; α_{M1}, α_{M2}, α_{M3}—intercept in model M1, M2 and M3, respectively, MW; $f_{M1}(Temp)$, $f_{M1}(Rad)$, $f_{M1}(Hour)$—one-dimensional additive functions of model M1, dependent on ambient air temperature, radiation, and hour of the day respectively, MW (see Figure 7a,b and Figure 8a); $f_{M1}(Wind,\ Temp)$—two-dimensional additive function of model M1, dependent on wind speed and ambient air temperature, MW (see Figure 8b); $f_{M2}(Temp)$, $f_{M2}(Hour)$—one-dimensional additive functions of model M2, dependent on ambient air temperature and hour of the day respectively, MW; $f_{M3}(Temp)$—one-dimensional additive function of model M3, dependent on ambient air temperature, MW; *Temp*—ambient air temperature (*Temp* = T_a), °C; *Rad*—solar radiation (*Rad* = I_r), W/m²; *Hour*—hour of the day (*Hour* = h), h and Wind—wind speed (*Wind* = W_s), m/s

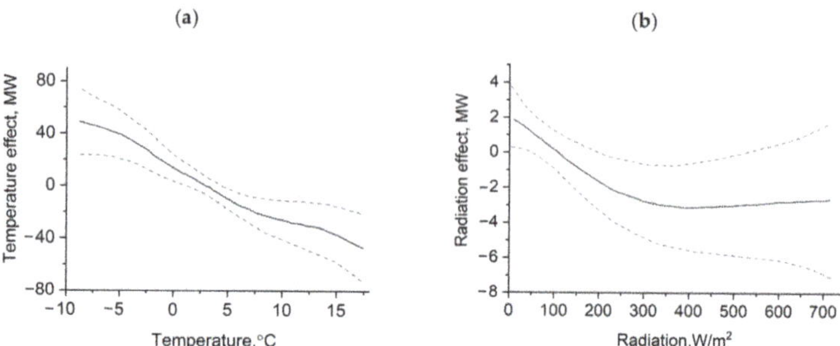

Figure 7. The additive effect on the heat load in the M1 model. The dashed line marks the 95% confidence interval (**a**) the air temperature effect $f(Temp)$ (**b**) solar radiation effect $f(Rad)$.

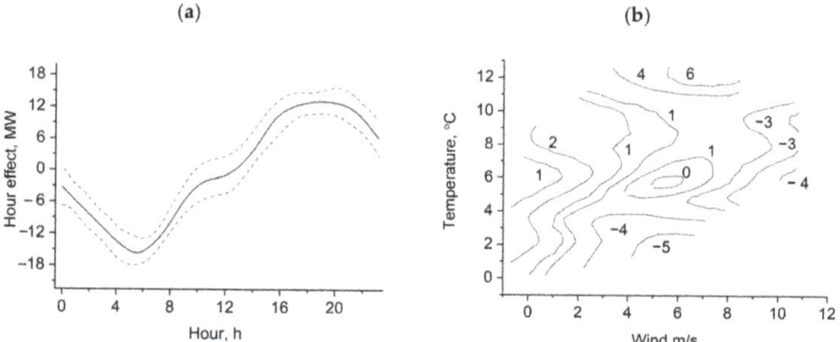

Figure 8. The additive effect on the heat load in the M1 model. The dashed line marks the 95% confidence interval (**a**) the hour effect $f(Hour)$ (**b**) two-dimensional air temperature and wind speed effect $f(Wind, Temp)$.

The detailed description of calculation Q_{M3} using M3 model (Equation (11)) is presented in Equation (12). The calculated coefficients are can be found in Table 2, where value of heat demand obtained from the model was also presented for the selected ambient air temperature:

$$Q_{M3} = \alpha_{M3} + f_{M3}(Temp) = \alpha_{M3} + \sum_{i=1}^{I=9} \beta_{T,M3,i} \cdot b_{T,M3,i}(Temp) \tag{12}$$

where: I—dimension of the spline basis (number of knots, $I = 9$); $\beta_{T,M3,i}$—the corresponding regression coefficient for M3 model (see Table 2) and $b_{T,M3,i}(Temp)$—the corresponding cubic-spline regression function for M3 model

Table 2. The corresponding regression coefficients together with intercept value in model M3 (for selected training dataset).

$\beta_{T,M3,i}$									α_{M3}, MW	Q_{M3}, MW (Temp = −0.7 °C)
i = 1	i = 2	i = 3	i = 4	i = 5	i = 6	i = 7	i = 8	i = 9		
8.55	−14.10	2.71	−6.03	1.46	4.73	2.03	27.86	−11.79	97.23	130.80

When calibrating the model (see Figure 5), input variables are fitted using smoothing terms. The signal from the heat meter on the output from the CHP to the DHN was used as the instantaneous heat load. Base functions are in the form of cubic spline functions. Other smooth functions such as thin plate regression spline or P-splines were examined. It was found that with a sufficiently high number of knots for splines, the type of function is of marginal importance. The additive term in M3 model, dependent on the ambient air temperature is shown in Equation (12). Other functions used in the remaining models (M2 and M1) are analogous. After the calibration stage, Equations (9)–(11) are used to find the forecasted heat load based on the new weather data.

In each model variant, a combination of input data and additive functions were selected to obtain the most accurate result in a day-ahead horizon. For example, it was observed that taking into account wind speed by using the two-dimensional additive function $f(Wind, Temp)$ gives a better fit. Table 3 presents the most relevant parameters used (as an argument for calculations within the *mgcv* package) [32].

Table 3. Input parameters for *mgcv* package used for building the predictive model.

Parameter	Description	Value
Family	The family object specifying the distribution and link to use in fitting.	Gaussian
Method	The smoothing parameter estimation method.	GCV (generalized cross validation)
Optimizer	The numerical method to optimize the smoothing parameter estimation criterion.	Newton
Smoothing functions	Indicating the smoothing basis to use.	Cubic regression splines

Figure 7 shows the impact of individual variables on the heat load for a sample training dataset in moving window (12 days) for the M1 model which contains four input variables. It can be clearly seen that as the ambient temperature together with solar irradiation increase, the values of the additive functions take lower values. For the variable representing hour, the daily variability of heat production is visible as a reduced power at night and increased in the evening (Figure 8a). The two-dimensional smoothing function, including the combined effect of wind speed and air temperature, can be seen in Figure 8b, where contour plot of additive term on the heat load is presented. The estimated value of the two dimensional additive function is marked on individual contour lines. One can notice that increasing the wind speed in the low temperature range leads to an increase in the heat load.

Table 4 contain some results from the fitting procedure of the M1 model, obtained for a selected training period. The table was generated as an output from *mgcv* package. It can be seen that the default maximum degrees of freedom for the smoothers used in the model are sufficient for all species, as the effective degrees of freedom (EDF) for all estimated smoothers are below their maximum possible value (k'). The p-value for the observed k-index is not significant. The k-index is a measure of the remaining pattern in the residuals, and the p-value is calculated based on the distribution of the k-index after randomizing the order of the residuals [34]. The data was fitted with GCV = 36.5 and $R^2 = 0.95$.

Table 4. Results of the fitting the M1 model.

Model Term	k'	EDF	k-Index	p-Value
$f(Temp)$	9.0	5.62	11.87	1.02×10^{-10}
$f(Rad)$	9.0	2.87	8.676	4.16×10^{-6}
$f(Wind, Temp)$	28.0	12.79	4.408	4.28×10^{-8}
$f(Hour)$	9.0	8.27	69.992	$<2 \times 10^{-16}$

EDF—effective degrees of freedom.

3.4. Effect of the Moving Window Size with Learning Data

In order to determine the optimal size of the sliding window, the effect of the number of days in the window was investigated. The analysis was based on the data from the entire heating season, simulating the operation of the investigated models day by day for different window sizes. The RMSE error was calculated for the test dataset (in the next day horizon), for forecasted as well as actual weather data. The results are illustrated in Figure 9. The graph shows that the optimal window size with the smallest corresponding RMSE error is 12 days. This window size allows the model to be adapted to the current DHN demand and end-users' behavior. The use of a time-shifting window potentially can lead to inaccurate forecasts in the next day's horizon (e.g., when the ambient air temperature is expected to rapidly decrease outside the range in the learning window). In that case, the forecasts will be extrapolated from smoothing functions.

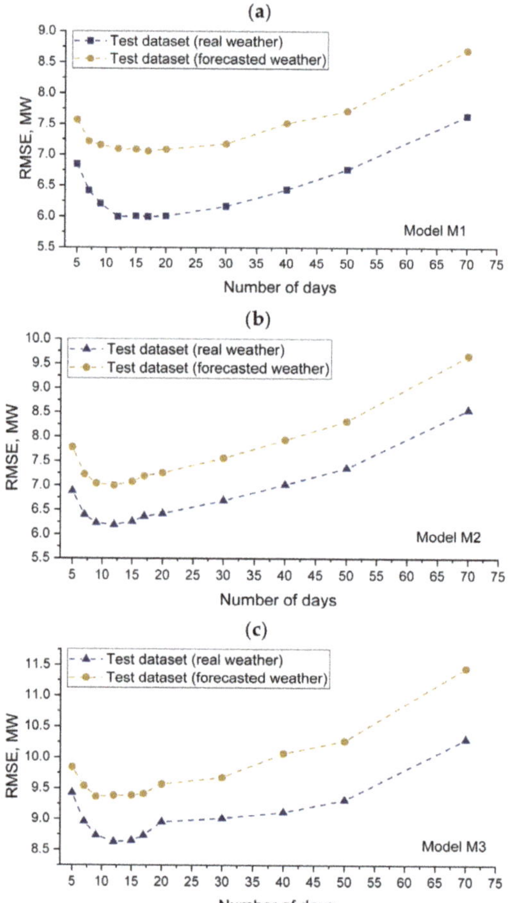

Figure 9. RMSE(root mean square error) depending on the size of the training window for (**a**) M1 model (**b**) M2 model (**c**) M3 model.

4. Results of the Heat Demand Model during the Heating Season

The obtained heat load models have been tested using the data coming from the DHS system during the heating season from November until March. The accuracy of the method was determined by means of a comparison of the forecasts with real values at the corresponding time. Figure 10 presents the time course of forecasts obtained by considered models along with the actual heat load for a selected period. It can be clearly seen that the M3 model is inaccurate compared with the M1 and M2 model. Including a variable representing the hour of the day significantly improves the results. This approach allows taking into account the daily profiles of heat production (e.g., morning and evening peaks).

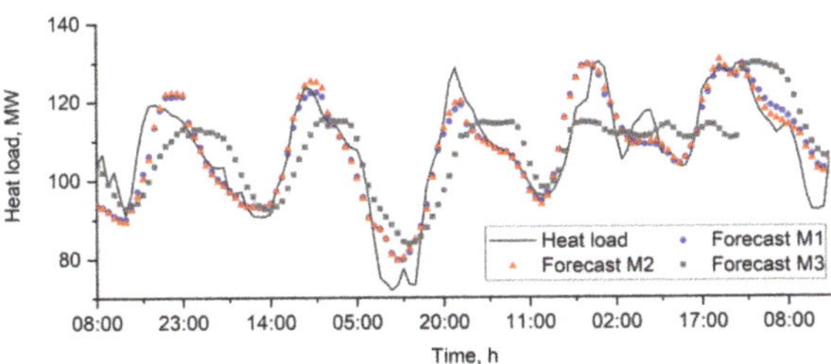

Figure 10. Actual and forecasted heat load over a several days.

From the point of view of optimizing the operation of the CHP plant, a high accuracy forecast for the forthcoming days is needed. It should be borne in mind that a model that fits well into the learning dataset may not give a good accuracy on the new data set. The Figure 11 shows the aggregated MAPE error for each model. The MAPE was calculated for the entire analyzed period (November–March). The metric was determined for the training dataset in a learning window (12 days) and test dataset (day-ahead horizon). In addition to weather data forecasts, simulation of model operation on real weather values was also included. This approach enables to assess the accuracy of the predictive model without weather forecast error. The M1 model gives a better fit to the training set because of including more input data (solar radiation and wind speed). A significant improvement can be observed while taking into account the hour of the day feature. This enables the daily seasonality to be taken into account.

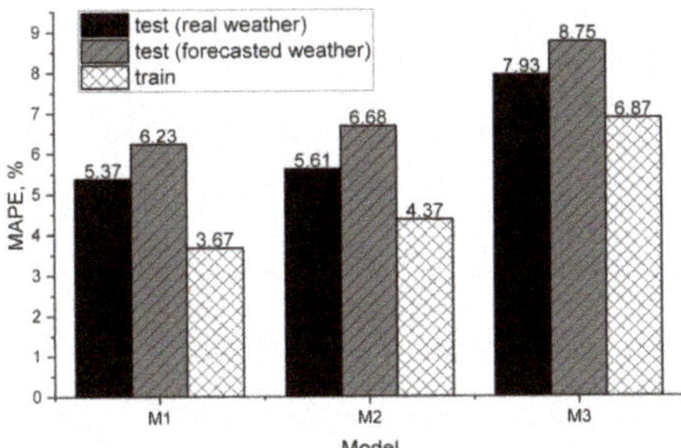

Figure 11. MAPE (mean absolute percentage error) for training and test dataset.

The box plot of absolute percentage errors for test dataset is presented in Figure 12. In the M1 and M2 model, the vast majority of hourly percentage errors are between 2 and 12%. The maximum observed errors exceed 20%. It can be seen that the M1 model gives slightly better results, particularly outliers are on a lower level. The Table 5 presents the results of the MAPE and RMSE metrics aggregated into individual months of the heating season (including M1 and M2 model). In the period from December to February, the models have similar accuracy. During transition periods such as November and March, where the heat demand is lower, the M1 model gives more accurate results. In these months, due to the relatively large daily amplitude of changes in ambient temperature as well as dynamic changes of mass flow rate in DHN, there are additional difficulties in forecasting. Standard deviation of the heat demand in these periods was greater than in typically winter months. It also should be noted that in these periods the weather forecast error increases. For example, the RMSE error of forecasts in ambient air temperature in January and February was approximately 1 °C. In March, RMSE increased to 1.5 °C.

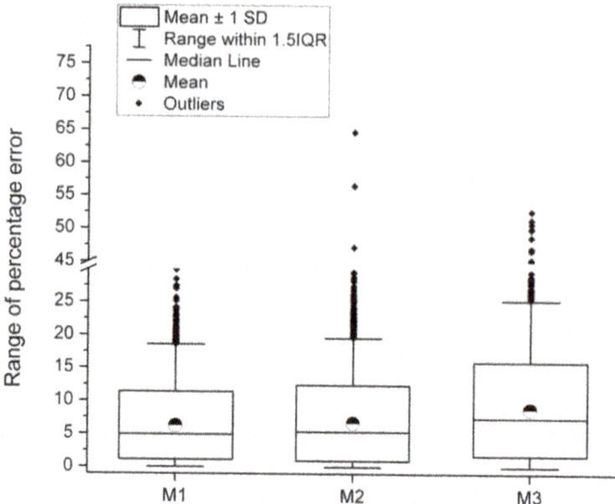

Figure 12. The boxplot of error for the test dataset (with forecasted weather).

Table 5. Statistical metrics in individual months of the M1 and M2 model.

Month	Mean Heat Load, MW	MAPE, %		RMSE, MW	
		Model M1	Model M2	Model M1	Model M2
November	110.1 (SD 18.7)	6.3	6.9	9.1	9.8
December	118.7 (SD 16.9)	5.2	5.4	7.6	7.9
January	118.3 (SD 16.1)	6.1	6.4	9.3	9.8
February	108.3 (SD 17.3)	5.9	6.0	7.9	7.9
March	96.5 (SD 22.4)	7.5	8.8	8.4	9.5

SD—Standard Deviation; MAPE—mean absolute percentage error; RMSE—root mean square error.

Including solar radiation in the learning data set allows to obtain better accuracy in March, when the influence of radiation in the thermal gains of buildings is greater. This can be observed in Figure 13. During periods of increased radiation over a day, the M1 model gives a better fit compared with the M2 model. The accuracy of two models is similar at night. The plot shows the hourly errors during several days from the end of March. During this period, the greatest instantaneous errors were observed. The time corresponds to the transient period between heating season and summer. Poor repeatability of heat production profile for similar climatic data was observed as a reason of inaccuracy. Therefore, the instantaneous errors exceed 20%, however the daily mean values are below 10%.

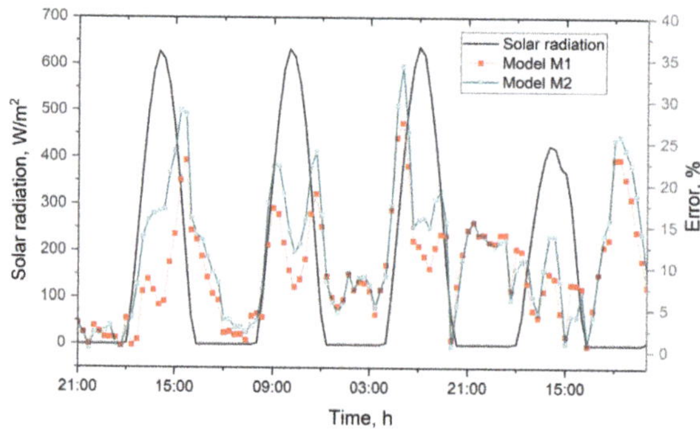

Figure 13. Solar radiation and absolute percentage error of the M1 and M2 model (end of March).

5. Conclusions

The paper presents the results of heat demand forecasting in the complex system supplied from a gas fired combined heat and power (CHP) plant with seven independent heat sources. The idea and results presented in the paper deal with the actual challenge of increasing energy efficiency during heat and electricity production. To maintain high production efficiency with reduced pollutants emission, high quality production forecasting is needed. For the purposes of optimizing heat production in cogeneration, hourly forecasts of the expected load demand in the forthcoming day are necessary. Accurate prediction enables efficient planning of heat production, taking into account the cooperation of the cogeneration heat sources with the district heating network. The main benefit comes from effective planning of electricity sales on the spot market, keeping the heat production at the required level. This is because the electricity production depends on the current heat load in the investigated system, which is a gas-fired combined heat and power plant. Electricity generation in gas-fired CHP plants allows reducing the emission of gaseous pollutants as CO_2, SO_x, and NO_x and stopping the emission of dust. The CO_2 emission level is around 50% lower in gas-fired power plants than coal-fired power plants. The main benefit of effective electricity generation planning in the gas-fired CHP plants is improving energy generation efficiency and reducing CO_2 emission level.

The generalized additive model (GAM) method was successfully used to build a predictive model based on weather data and a variable representing an hour of the day. The heat load profile can change over time because of thermo-modernization of buildings, changing the regulation of heating nodes or new connections to the heating network. For this reason, it is important to calibrate the model properly and continuously. In presented paper the application of the adaptive training of the model using moving window was investigated. The results confirmed that the last 12 days give the opportunity to take into account the current conditions of DHN operation and heat consumption behavior by end users. The moving window was adopted in all three variants of the heat demand forecasting models. The model M1 is supplied with ambient temperature, radiation and wind speed. Model M2 was additionally supplied with variable representing the hour of the day, and model M3 only consider ambient air temperature. The results of model M1, including additional weather data such as irradiation and wind speed gives the best results, particularly between the winter and summer period when high fluctuations of radiation during a whole day exist. It should be noted that the model is also burdened with independent factors such as weather forecast error and uncertainty in measuring the thermal power at the DHN supply side (about 1.5–2%).

Author Contributions: Conceptualization, M.B. and P.M.; methodology, M.B. and P.M.; software, M.B.; validation, M.B., and P.M.; investigation, M.B. and P.M.; resources, M.B.; writing—original draft preparation, M.B. and P.M.; writing—review and editing, P.M.; visualization, M.B.; supervision, P.M. All authors have read and agreed to the published version of the manuscript.

Funding: This research was funded by MEiN subsidy no. 16.16.130.942 (Faculty of Mechanical Engineering and Robotics, AGH UST).

Institutional Review Board Statement: Not applicable.

Informed Consent Statement: Not applicable.

Data Availability Statement: Restrictions apply to the availability of these data. Data was obtained from PGE Energia Ciepła S.A. and are available from the authors with the permission of PGE Energia Ciepła S.A.

Acknowledgments: The research was carried out thanks to the possibility of participating in the Implementation Doctorate Programme at the AGH Doctoral School.

Conflicts of Interest: The authors declare no conflict of interest.

References

1. Żymełka, P.; Szega, M.; Madejski, P. Techno-Economic Optimization of Electricity and Heat Production in Gas-Fired CHP Plant with Heat Accumulator. *J. Energy Resour. Technol.* **2020**, *142*, 022101. [CrossRef]
2. Fang, T.; Lahdelma, R. Optimization of combined heat and power production with heat storage based on sliding time window method. *Appl. Energy* **2016**, *162*, 723–732. [CrossRef]
3. Wang, J.; You, S.; Zong, Y.; Cai, H.; Træholt, C.; Dong, Z.Y. Investigation of real-time flexibility of combined heat and power plants in district heating applications. *Appl. Energy* **2019**, *237*, 196–209. [CrossRef]
4. Sameti, M.; Haghighat, F. Optimization of 4th generation distributed district heating system: Design and planning of combined heat and power. *Renew. Energy* **2019**, *130*, 371–387. [CrossRef]
5. Schweiger, G.; Kuttin, F.; Posch, A. District Heating Systems: An Analysis of Strengths, Weaknesses, Opportunities, and Threats of the 4GDH. *Energies* **2019**, *12*, 4748. [CrossRef]
6. Madejski, P.; Żymełka, P.; Węzik, R.; Kubiczek, H. Gas fired plant modeling for monitoring and optimization of electricity and heat production. *J. Power Technol.* **2017**, *97*, 455–462.
7. Idowu, S.; Saguna, S.; Åhlund, C.; Schelén, O. Applied machine learning: Forecasting heat load in district heating system. *Energy Build.* **2016**, *133*, 478–488. [CrossRef]
8. Spoladore, A.; Borelli, D.; Devia, F.; Mora, F.; Schenone, C. Model for forecasting residential heat demand based on natural gas consumption and energy performance indicators. *Appl. Energy* **2016**, *182*, 488–499. [CrossRef]
9. Nigitza, T.; Gölles, M. A generally applicable, simple and adaptive forecasting method for the short-term heat load of consumers. *Appl. Energy* **2019**, *241*, 73–81. [CrossRef]
10. Mosavi, A.; Bahmani, A. Energy Consumption Prediction Using Machine Learning. *Electr. Eng.* **2019**, 2019030131.
11. Bourdeau, M.; Zhai, X.Q.; Nefzaoui, E.; Guo, X.; Chatellier, P. Modeling and forecasting building energy consumption: A review of data-driven techniques. *Sustain. Cities Soc.* **2019**, *48*, 101533. [CrossRef]
12. Dotzauer, E. Simple model for prediction of loads in district-heating systems. *Appl. Energy* **2002**, *73*, 277–284. [CrossRef]
13. Baltputnis, K.; Petrichenko, R.; Sobolevsky, D. Heating Demand Forecasting with Multiple Regression: Model Setup and Case Study. In Proceedings of the IEEE 6th Workshop on Advances in Information, Electronic and Electrical Engineering (AIEEE), Vilnius, Lithuania, 8–10 November 2018; pp. 1–5. [CrossRef]
14. Grosswindhager, S.; Voigt, A.; Kozek, M. Online Short-Term Forecast of System Heat Load in District Heating Networks. In Proceedings of the 31st International Symposium on Forecasting, Prague, Czech Republic, 26–29 June 2011.
15. Fang, T.; Lahdelma, R. Evaluation of a multiple linear regression model and SARIMA model in forecasting heat demand for district heating system. *Appl. Energy* **2016**, *179*, 544–552. [CrossRef]
16. Wojdyga, K. Predicting Heat Demand for a District Heating Systems. *Int. J. Energy Power Eng.* **2014**, *3*, 237. [CrossRef]
17. Saloux, E.; Candanedo, J.A. Forecasting District Heating Demand using Machine Learning Algorithms. *Energy Procedia* **2018**, *149*, 59–68. [CrossRef]
18. Bandyopadhyay, S.; Hazra, J.; Kalyanaraman, S. A machine learning based heating and cooling load forecasting approach for DHC networks. In Proceedings of the 2018 IEEE Power & Energy Society Innovative Smart Grid Technologies Conference (ISGT), Washington, DC, USA, 19–22 February 2018; pp. 1–5.
19. Suryanarayana, G.; Lago, J.; Geysen, D.; Aleksiejuk, P.; Johansson, C. Thermal load forecasting in district heating networks using deep learning and advanced feature selection methods. *Energy* **2018**, *157*, 141–149. [CrossRef]
20. Le, L.T.; Nguyen, H.; Zhou, J.; Dou, J.; Moayedi, H. Estimating the Heating Load of Buildings for Smart City Planning Using a Novel Artificial Intelligence Technique PSO-XGBoost. *Appl. Sci.* **2019**, *9*, 2714. [CrossRef]
21. Dahl, M.; Brun, A.; Kirsebom, O.S.; Andresen, G.B. Improving Short-Term Heat Load Forecasts with Calendar and Holiday Data. *Energies* **2018**, *11*, 1678. [CrossRef]
22. Geysen, D.; De Somer, O.; Johansson, C.; Brage, J.; Vanhoudt, D. Operational thermal load forecasting in district heating networks using machine learning and expert advice. *Energy Build.* **2018**, *162*, 144–153. [CrossRef]
23. Wood, S. *Generalized Additive Models: An Introduction with R*; Chapman & Hall: London, UK, 2006.
24. Krstonijević, S. Generalized Additive Model for Electricity Load Prediction in R. *Zenodo* **2020**. [CrossRef]
25. Kim, H.; Kim, J. Short term electricity demand forecasting in south korea with generalized additive models. *J. Korean Data Inf. Sci. Soc.* **2018**, *29*, 1299–1307. [CrossRef]
26. Sigauke, C. Forecasting medium-term electricity demand in a South African electric power supply system. *J. Energy South. Afr.* **2017**, *28*, 54. [CrossRef]
27. Pathak, N.; Ba, A.; Ploennigs, J.; Roy, N. Forecasting Gas Usage for Big Buildings Using Generalized Additive Models and Deep Learning. In Proceedings of the 2018 IEEE International Conference on Smart Computing (SMARTCOMP), Taormina, Italy, 18–20 June 2018; pp. 203–210.
28. Khamma, T.R.; Zhang, Y.; Guerrier, S.; Boubekri, M. Generalized additive models: An efficient method for short-term energy prediction in office buildings. *Energy* **2020**, *213*, 118834. [CrossRef]
29. Sandou, G.; Font, S.; Tebbani, S.; Hiret, A.; Mondon, C. Optimization and control of a district heating network. *IFAC Proc. Vol.* **2005**, *38*, 397–402. [CrossRef]
30. Hastie, T.; Tibshirani, R. *Generalized Additive Models*; Chapman & Hall/CRC: Boca Raton, FL, USA, 1990.

31. Perperoglou, A.; Sauerbrei, W.; Abrahamowicz, M.; Schmid, M. A review of spline function procedures in R. *BMC Med Res. Methodol.* **2019**, *19*, 1–16. [CrossRef]
32. Wood, S. mgcv: GAMs and Generalized Ridge Regression for R. *R News* **2001**, *1*, 20–25.
33. Craven, P.; Wahba, G. Smoothing noisy data with spline functions. *Numer. Math.* **1978**, *31*, 377–403. [CrossRef]
34. Pedersen, E.J.; Miller, D.L.; Simpson, G.L.; Ross, N. Hierarchical generalized additive models in ecology: An introduction with mgcv. *PeerJ* **2019**, *7*, e6876. [CrossRef] [PubMed]

MDPI
St. Alban-Anlage 66
4052 Basel
Switzerland
Tel. +41 61 683 77 34
Fax +41 61 302 89 18
www.mdpi.com

Energies Editorial Office
E-mail: energies@mdpi.com
www.mdpi.com/journal/energies